Research with Diverse Groups

POCKET GUIDES TO
SOCIAL WORK RESEARCH METHODS

Series Editor
Tony Tripodi, DSW
Professor Emeritus, Ohio State University

Determining Sample Size: Balancing Power, Precision, and Practicality
Patrick Dattalo

Preparing Research Articles
Bruce A. Thyer

Systematic Reviews and Meta-Analysis
Julia H. Littell, Jacqueline Corcoran, and Vijayan Pillai

Historical Research
Elizabeth Ann Danto

Confirmatory Factor Analysis
Donna Harrington

Randomized Controlled Trials: Design and Implementation for Community-Based Psychosocial Interventions
Phyllis Solomon, Mary M. Cavanaugh, and Jeffrey Draine

Needs Assessment
David Royse, Michele Staton-Tindall, Karen Badger, and J. Matthew Webster

Multiple Regression with Discrete Dependent Variables
John G. Orme and Terri Combs-Orme

Developing Cross-Cultural Measurement
Thanh V. Tran

Intervention Research: Developing Social Programs
Mark W. Fraser, Jack M. Richman, Maeda J. Galinsky, and Steven H. Day

Developing and Validating Rapid Assessment Instruments
Neil Abell, David W. Springer, and Akihito Kamata

Clinical Data-Mining: Integrating Practice and Research
Irwin Epstein

Strategies to Approximate Random Sampling and Assignment
Patrick Dattalo

Analyzing Single System Design Data
William R. Nugent

Survival Analysis
Shenyang Guo

The Dissertation: From Beginning to End
Peter Lyons and Howard J. Doueck

Cross-Cultural Research
Jorge Delva, Paula Allen-Meares, and Sandra L. Momper

Secondary Data Analysis
Thomas P. Vartanian

Narrative Inquiry
Kathleen Wells

Structural Equation Modeling
Natasha K. Bowen and Shenyang Guo

Finding and Evaluating Evidence: Systematic Reviews and Evidence-Based Practice
Denise E. Bronson and Tamara S. Davis

Policy Creation and Evaluation: Understanding Welfare Reform in the United States
Richard Hoefer

Grounded Theory
Julianne S. Oktay

Systematic Synthesis of Qualitative Research
Michael Saini and Aron Shlonsky

Quasi-Experimental Research Designs
Bruce A. Thyer

Conducting Research in Juvenile and Criminal Justice Settings
Michael G. Vaughn, Carrie Pettus-Davis, and Jeffrey J. Shook

Qualitative Methods for Practice Research
Jeffrey Longhofer, Jerry Floersch, and Janet Hoy

Analysis of Multiple Dependent Variables
Patrick Dattalo

Culturally Competent Research: Using Ethnography as a Meta-Framework
Mo Yee Lee and Amy Zaharlick

Using Complexity Theory for Research and Program Evaluation
Michael Wolf-Branigin

Basic Statistics in Multivariate Analysis
Karen A. Randolph and Laura L. Myers

Research with Diverse Groups: Research Designs and Multivariate Latent Modeling for Equivalence
Antoinette Y. Farmer and G. Lawrence Farmer

ANTOINETTE Y. FARMER
G. LAWRENCE FARMER

Research with Diverse Groups

Research Designs and Multivariate Latent Modeling for Equivalence

Oxford University Press is a department of the University of Oxford. It furthers the University's objective of excellence in research, scholarship, and education by publishing worldwide.

Oxford New York
Auckland Cape Town Dar es Salaam Hong Kong Karachi
Kuala Lumpur Madrid Melbourne Mexico City Nairobi
New Delhi Shanghai Taipei Toronto

With offices in
Argentina Austria Brazil Chile Czech Republic France Greece
Guatemala Hungary Italy Japan Poland Portugal Singapore
South Korea Switzerland Thailand Turkey Ukraine Vietnam

Published in the United States of America by
Oxford University Press
198 Madison Avenue, New York, NY 10016

Library of Congress Cataloging-in-Publication Data
Rodgers-Farmer, Antoinette Y.
Research with diverse groups : research designs and multivariate latent modeling for equivalence / by Antoinette Y. Farmer and G. Lawrence Farmer.
p. cm.—(Pocket guides to social work research methods)
Includes bibliographical references and index.
ISBN 978-0-19-991436-4 (alk. paper)
1. Social service—Research. 2. Multiculturalism—Research.
3. Multivariate analysis. I. Farmer, G. Lawrence. II. Title.
HV11.R526 2014
361.4072—dc23
2013037910

1 3 5 7 9 8 6 4 2
Printed in the United States of America
on acid-free paper

Contents

Acknowledgments

We want to thank Dr. Tony Tripodi for providing us with the opportunity to write this book as part of the Oxford University Press Pocket Guide to Social Work Research Methods Series. We are extremely grateful for the support that we received from Nicholas Liu, Agnes Bannigan, the Oxford University Press staff, and the anonymous reviewers for their thorough comments. I, Antoinette Y. Farmer, would like to acknowledge all the following persons who have had an influence on my intellectual development: Dr. Gary F. Koeske (dissertation chair), Father Hodge and the late Mrs. Martha Hodge (my long-time friends), my parents, and my grandmother. I especially want to thank my administrative assistant, Elizabeth Koechlein. Her assistance with the preparation of this book was invaluable. Both Lawrence and I want to thank our children, Bradford, Evan, and Elliott, for providing us with the emotional support we so much needed while we were writing this book.

Research with Diverse Groups

1

Introduction

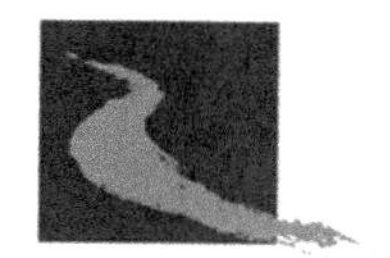

The purpose of this book is to illustrate how to achieve research-design equivalence across the diverse groups in one's study. Groups can be diverse with respect to individual characteristics; for example, in gender, ethnicity, and sexual orientation; or in the characteristics of the settings in which the group resides; for example, urban and suburban communities. In this text we are assuming that individual and contextual aspects of persons' diversity shape their experiences, understanding, and expression of the social phenomenon researchers are investigating.

Research-design equivalence refers to the use of processes and procedures that ensure accurate representation of the phenomenon under investigation across diverse groups. For example, consider a study that has the goal of determining what role the local recreational institutions play in shaping urban and rural teenagers' experiences of social exclusion. The researcher will need to tailor every aspect of the study's design in a way that maximizes the internal and external validity of the study for each of the two groups. For example, the measure of social exclusion must be designed to fit the unique context in which each group of youth resides. The measure must have group-specific relevance and meaning. Even if the researcher is seeking to develop an understanding of the social exclusion that is not group-specific, the measure must also have comparable meaning across the groups. The researcher will have to ensure that the measure has comparable reliability and validity across the diverse groups of youth. The measures used during the data collection phase are just one aspect of the study's design that requires attention to the comparability across diverse samples. The concern about the research-design equivalence begins with the formulation of the research question and ends with the interpretation and reporting of findings.

The accumulation of empirical knowledge across studies, which is the hallmark of the scientific enterprise, is not possible if research-design equivalence across studies is not achieved. Without research-design equivalence, biased conclusions will be drawn by those seeking to synthesize findings across studies (van Herk, Poortinga, & Verhallen, 2005). An ongoing challenge for those seeking to ground social work practice in an evidence-based practice framework is tackling the problems in establishing research-design equivalence across studies when conducting and trying to apply findings from systematic reviews and meta-analyses (Kriston, 2013; Nugent, 2012).

This book intends to describe the unique methodological issues that social work researchers face when conducting research with diverse groups, and to provide some guidance for how to address these issues. It is meant to supplement writings on how to conduct cross-culture research found in the works of Liamputtong (2010), Matsumoto and van de Vijver (2011), and F. van de Vijver and K. Leung (1997). Those involved in conducting cross-cultural research are encouraged to consult these writings. Additionally, this book is designed to serve as supplement to standard research-methods texts used in social work doctoral programs that too often do not mention the methodological issues researchers face when conducting research with diverse groups.

This book is written for social work doctoral students who are interested in conducting research with diverse groups. This book should also be of interest to doctoral students in other professions, such as psychology, nursing, education, and public health, as they, too, may be interested in conducting research with diverse groups. This book builds on an understanding of research design (Thyer, 2010; Viswanathan, 2005), classical test theory (Carmines & Zeller, 1979; Nugent & Hankins, 1992), and factor analysis (Pett, Lackey, & Sullivan, 2003). It is assumed that the readers have an understanding of how to apply the concepts of internal and external validity to various research designs and the elements of a research design. Readers should understand how reliability, validity, and measurement error are conceptualized. An understanding of the application of exploratory and confirmatory factor analysis as analytical strategies for the evaluation of the validity of a measure needs to be understood. Confirmatory factor analysis as a tool for the evaluation of a measure's equivalency across groups will be demonstrated using Mplus 7 (Muthén & Muthén, 1998–2012). On the developer's website (StatModl.com), readers can learn how to use Mplus7 and download a fully functioning demonstration version. The program is available for Windows, Linux, and Mac OS X.

DIVERSITY: ITS IMPLICATIONS FOR ESTABLISHING EQUIVALENCE

The United States population is quite diverse—in terms of gender, race or ethnicity, ability, sexual orientation, socioeconomic status, and culture. This diversity has implications for establishing equivalence for all aspects of a study's design. It has impacts on problem formulation, sampling, and measurement selection. The implications of diversity on these aspects of the research process will be discussed below.

Problem Formulation

Various aspects of individuals' diversity will affect the way they conceptualize the phenomena under investigation (Carter-Black & Kayama, 2011). For example, Carter-Black and Kayama (2011) found that socio-economic status shaped the way racism was experienced and conceptualized by two African American women who shared several social markers of diversity (ethnicity, gender, and historical and regional contexts). The findings from this study illustrate how individuals' diversity needs to be taken into account when formulating the research questions.

Sampling Equivalence

Diversity has implications for representative sampling. Obtaining a sample that is representative of the diversity found in the population under study may be a daunting task due to diverse subgroups found in the population (Knight, Roosa, & Umaña-Taylor, 2009). For example, there are 566 federally recognized Native American and Alaska Native tribes and villages (U.S. Department of the Interior, 2013). The Native American and Alaska Native populations consist of distinct tribes and ethnic groups. Researchers studying Native American and Alaska Natives may find it difficult to get a representative sample of Native Americans and Alaska Natives due to diversity within this group (Knight et al., 2009), as they would need to ensure the appropriate representativeness of the sample by tribe and ethnic group.

Measurement Selection

Establishing measurement equivalence requires social work researchers to demonstrate that they are measuring the same construct across groups in their study. Demonstrating that one is measuring the same construct across groups is challenging, in part because persons' experiences shape their perceptions and understanding of the phenomenon under study. For example, research looking at depression in African American males has to take into account how gender, African American culture, racism (Sean, 2005), and the complex interaction among these variables affect their expression of depressive symptoms. In examining how middle-aged African American men expressed symptoms of depression, Bryant-Bedell and Waite (2010) found that they described their symptoms as "being in a funk," which was later identified as depression by a healthcare professional. Variations in response styles across the groups being compared makes it difficult to establish equivalence and affects researchers' ability to interpret the results (Knight et al., 2009). Research has demonstrated that less acculturated Hispanics have a more extreme response style than highly acculturated Hispanics (Marin, Gamba, & Marin, 1992). Differences in the measurement error across groups will make it difficult to assess structural equivalence. Establishing structural equivalence is vital when researchers want to test the assumption that the construct has the same dimensionality across groups as expected by theory (Byrne & van de Vijver, 2010). Differing sets of beliefs about a behavior may make trying to establish measure equivalence difficult. For example, Knight et al. (2009) suggest that an item on a depression measure could have less variability in responses in one group because of the group members' religious beliefs.

We have chosen two diverse groups to illustrate the types of diversity that one must take into consideration when conducting research with such groups. Additionally, we discussed the implications of these types of diversity for establishing equivalence. Our not illustrating the types of diversity within other diverse groups within the United States does not negate the need to include these diverse groups in one's research. Although we have not described all of the diverse groups or aspects of diversity within the United States, the methodological and statistical approaches we describe in this book can be used with these diverse groups as well.

African Americans

According to the 2010 U.S. Census Bureau, African Americans make up 12.6% of the total U.S. population (U.S. Census Bureau, 2011a). Not included in the above 12.6% is an additional 1% of African Americans who identified themselves as African American in combination with one or more other ethnic groups (U.S. Census Bureau, 2011a). Therefore, 13.6% of the U.S. population is African Americans, either alone or in combination with one or more other ethnic groups. It is estimated that by 2060, African Americans will make up 18.1% of the total U.S. population (U.S. Census Bureau, 2012). The definition of "African American" used in the 2010 Census refers to having origins in any Black racial groups. Included in this category are persons who consider themselves sub-Saharan Africans (except Sudanese and Cape Verdeans) and Afro-Caribbean. Individuals from North Africa are not defined as African American but as White. The way in which African Americans are defined by the U.S. Census Bureau indicates the African American population is diverse. The majority of African Americans live in the South (55%), with six southern states (Mississippi, Louisiana, Georgia, Maryland, South Carolina, and Alabama) and the District of the Columbia having the largest number of African Americans: 37%, 33%, 32%, 31%, 29%, 27%, and 52%, respectively (U.S. Census Bureau, 2011a). In 2010, the median household income of African Americans was $32,584, compared with $50,046 for all U.S. families (U.S. Census Bureau, 2010).

Hispanics

Currently, there are 50.5 million Hispanics or persons of Latino origin in the United States, comprising 16% of the population (U.S. Census Bureau, 2011b). These numbers do not include the 3.1 million

U.S. citizens who live on the island of Puerto Rico (U.S Census Bureau, 2011b). It is estimated that by 2060, Hispanics will make up 30.6% of the total U.S. population (U.S. Census Bureau, 2012). The Census Bureau (2011b) defines "Hispanic" or "Latino" persons, regardless of race, as Cuban, Mexican, Puerto Rican, South or Central American, or of other Spanish culture or origin. "Hispanic" is considered to be an ethnicity and not a race. Persons who identify themselves as Hispanic are asked to specify their race. Between 2000 and 2010, the Hispanic population increased by 15.2 million (U.S. Census Bureau, 2011b). The Hispanic groups who experienced the most growth during this time period were Mexicans, Puerto Ricans, and Cubans. The majority of the Hispanic population resides in the West, with the greatest concentration of this population living in New Mexico (46.3%), followed by Texas and California (37.6% in each state; U. S. Census Bureau, 2011b). Hispanics are more likely to live in these states as they are along the border with Mexico. Although most Hispanics live in the West, there is geographic diversity associated with various Hispanic groups. For example, Mexicans are more likely to reside in Texas; Cubans are more likely to reside in Florida; and Salvadorans are more likely to reside in Maryland (U.S. Census Bureau, 2011b). The groups of persons who are considered to be Hispanic differ in terms of country of origin, customs, and variations in Spanish spoken. The median household income for Hispanics in 2010 was $40,165, compared with $50,046 for all U.S. families (U.S. Census Bureau, 2010).

Both the African American and the Hispanic population in the U.S. also have within-group diversity (subpopulations within the diverse group). Within-group diversity has implications for establishing conceptual equivalence. Therefore, it is important that within-group diversity be accounted for when conducting research with these groups. Several methodological and statistical approaches will be discussed in this book that social work researchers can use to account for within-group diversity. When conducting research with Hispanics, researchers need to assess their acculturation and immigration status, as these factors may have implications for establishing structural equivalence. These groups are also diverse in terms of which parts of the United States they live in. Therefore, geographic diversity needs to be taken into consideration when establishing equivalence across groups.

The aspects of diversity of these two groups described earlier would be considered *observable diversity* (i.e., manifest variables). Additionally, there are aspects of diversity associated with these groups that are considered *unobservable diversity* (i.e., latent variables), which also have implications for how one operationalizes the constructs being measured in one's study. For example, sexual expression is not an observable variable, but it does have an effect on how the person conceptualizes masculinity and femininity. Advances in the estimation of latent class models make it possible for researchers to identify and incorporate latent aspects of diversity into the research process (Bollen, 2002; Moisio, 2004; Muthén, 2002).

The rapidly changing demographics of the United States; the heightened need to understand why health, economic, and educational disparities exist between certain segments of the population; and the need to develop effective interventions for various groups continue to fuel the need to conduct research with diverse groups. Therefore, social workers need to be well equipped with the skills needed to design methodologically sound studies for research with diverse groups. Designing such studies requires that social workers pay greater attention to ensuring that equivalence has been established at all phases of the research process. Ideally, if greater attention is paid to establishing equivalence at each phase of the research process, then it is more likely that the results can be attributed to true group differences.

THE NEED TO CONSIDER CONTEXTUAL FACTORS WHEN ESTABLISHING MEASUREMENT EQUIVALENCE

Oftentimes social work researchers have used the ecological perspective to guide their research. Using the perspective, they examine how an individual's characteristics and environment (e.g., family, school, community, and peer group) affect the phenomenon under investigation. These five domains (individual, family, peer, school, and community) can differ across the diverse groups being studied. For example, in conducting a study examining the effects of parenting practices on African and European American adolescents, one could have African American adolescents who live in low socioeconomic environments and European American adolescents who live in middle-class socioeconomic environments. Contextual factors not only have a direct effect on the outcome variable under investigation but also influence the reliability and validity of its measurement. Contextual factors can potentially effect the equivalency of factor loadings and error variances across groups.

ORGANIZATION OF THE BOOK

This book focuses on the need to establish equivalence as it relates to various phases of the research process: problem formulation, research design, sampling, measurement selection, data collection, and data analysis. Any phase of the research process can result in erroneous conclusions if equivalence is not established, so researchers should be continually mindful of this issue when designing their studies. At the problem-formulation stage, researchers should establish construct equivalence. Establishing construct equivalence could involve assessing for conceptual/configural equivalence, functional (concurrent and/or predictive validity) equivalence, metric and scalar equivalence. When developing the sampling frame, researchers need to establish sampling equivalence. Establishing sampling equivalence reduces the threat of a composition effect. A *composition effect* occurs when certain individuals have a higher probability of being included in the sample than others, especially when stratified sampling is used to select one group of participants, and convenience sampling is used to select another. Establishing conceptual equivalence is critical not only at the problem-formulation stage but also at the measurement-selection phase, as it is important that the construct of interest has the same meaning across groups. Before collecting the data, researchers should ensure procedural equivalence. *Procedural equivalence* refers to ensuring consistency across groups in the way the surveys are administered, the timing of the surveys' administration, the conditions under which the surveys are administered, and the mode of data collection (Schaffer & Riordan, 2003). In analyzing the data, researchers should establish measurement and structural equivalence.

A discussion of all types of measurement equivalence is beyond the scope of this book, but readers who are interested in learning more about the various types of measurement equivalence not discussed here can refer to Johnson (2006). We do, however, focus on the seven types of measurement equivalence discussed by Milfont and Fischer (2010), because they are particularly important when conducting research with diverse groups. These seven types of equivalence can be classified into two categories: measurement and structural. *Measurement equivalence* focuses on establishing equivalence by examining how items function across groups. Measurement equivalence can be assessed by examining factor loadings, item intercepts, and error variances across groups (these are considered to be observed variables; Milfont & Fischer, 2010). *Structural equivalence* refers to establishing the theoretical structure of the measure across groups (Byrne & Watkins, 2003). This type of equivalence can be assessed by examining factor variance, factor covariance, and factor means across groups (these are considered to be unobserved or latent variables).

The types of measurement equivalence we discuss are configural, metric, scalar, and error (i.e., covariance). Assessing configural equivalence helps researchers determine whether the groups differ on how they conceptualize the variable of interest (Adamsons & Buehler, 2007). Testing for metric equivalence is important because, if present, it demonstrates that the groups under investigation are responding to a given item on the measure in the same manner (Milfont & Fischer, 2010). When conducting research with diverse groups, it is important to establish scalar equivalence (also referred to as "strong metric equivalence") to ensure that the cutoff scores of a particular measure are the same for all the groups in the study. If the cutoff scores are different, that would have implications for how interventions are developed for each of the groups under investigation. Establishing error equivalence (also referred to as "strict metric equivalence") is important because measures cannot operate equivalently across groups if the error variances differ (DeShon, 2004). When one or more of the above types of measurement equivalence (configural, metric, scalar, or error) have been established, the researchers can test for structural equivalence by comparing the groups based on their factor variances, factor covariances, and factor means (Meredith, 1993). Comparing the groups based on the factor means allows researchers to test for differences in the underlying constructs between the groups, whereas comparing the factor variances allows researchers to examine whether the range of responses varies between groups (Adamsons & Buehler, 2007). For Vandenberg and Lance (2000), demonstrating both the measurement and structural equivalence of a measure is just as important as demonstrating the reliability and validity of the measure. Establishing structural equivalence is vital when researchers want to test the assumption that the construct has the same dimensionality across groups as expected by theory (Byrne & van de Vijver, 2010).

Social work researchers should understand the various statistical methods that can be used to establish both measurement and structural equivalence. In this book, we focus on how structural equation modeling (SEM) can be used within the framework of a confirmatory factor analysis (CFA) model to establish the above types of equivalence. Specifically, we describe how SEM can be used to establish equivalence in the context of multi-group confirmatory factor analysis (MG-CFA).

In Chapter 2, we describe the major phases of the research process in which attention must be given to achieve research-design equivalence. Using the research process as the conceptual framework, we highlight the major phases of the research process for which equivalence must be established: problem formulation, research design, sampling, measurement selection, and data collection (except for the data analysis phase, which will be discussed in Chapter 3). We discuss the methodological issues that may result in there being non-equivalence across the groups, resulting in erroneous conclusions about the findings obtained as well as the strategies that can be used to address these issues that will produce equivalence across the groups. In addition, we discuss the importance of ruling out alternative explanations for one's findings as the result of using non-experimental research designs. To illustrate the concepts discussed in Chapter 2, we will present a hypothetical case in Chapter 4.

Chapter 3 is devoted to discussing the seven types of equivalence that Milfont and Fischer (2010) argued are important to establish when conducting research with diverse groups. We also discuss the rationale for why each type of measurement equivalence needs to be established. Specifically, we focus on establishing measurement equivalence across groups when the groups have been identified based on directly observable characteristics (e.g., gender, ethnicity) or manifest variables. An MG-CFA is described as the statistical approach to test each of the seven types of equivalence that need to be established. Using data from the National Longitudinal Study of 1988 (NELS:88), an MG-CFA was conducted using SEM to illustrate the seven types of equivalence. The results are presented and discussed. The Mplus syntax for the SEM analyses are provided. A write-up of how each of these analyses would be reported in a publication is also presented.

Chapter 4 presents a hypothetical case demonstrating how equivalence can be established at each phase of the research process. In addition, readers are introduced to how descriptive statistics can be reviewed to determine if non-equivalence or equivalence initially exists across the groups for the variable of interest, prior to conducting an MG-CFA to establish the seven types of equivalence described in this book.

Chapter 5 discusses the use of qualitative methods to establish measurement equivalence; the challenges of conducting research to establish equivalence using national datasets; and directions for social work education at the doctoral level. Additionally, the contributions of this book to research methods are highlighted.

SIGNIFICANCE FOR SOCIAL WORK

Generally speaking, most studies in the field of social work can be considered comparative research. These studies do not usually compare individuals from different cultures; rather, they tend to compare two or more groups of people of different genders, ethnicities, or sexual orientations. When conducting comparative research, social workers must be knowledgeable about the best ways to ensure that the measures used in their studies are sensitive enough to detect true differences between groups. As Gregorich (2006) succinctly stated,

> Defensible use of self-reports in quantitative comparative research requires not only that the measured constructs have the same meaning across groups, but also that the group comparisons of the sample estimates (e.g., means and variances) reflect true group differences and are not contaminated by group-specific attributes that are unrelated to the construct of interest. (p. S78)

Because social work research commonly involves diverse groups, it is important that researchers know how to establish at least the seven types of equivalence discussed in this book. It should be noted that even if metric, conceptual, and other types of equivalence have been established using appropriate statistical techniques, a study's inferences may still be invalid if the design, sampling process, or survey administration are flawed. Therefore, it is important that equivalence be established at all phases of the research process prior to data analysis. In doing so, the field will be able to produce more valid and reliable findings so that more accurate policies and effective interventions can be developed.

2

Research-Design Equivalence

OVERVIEW

In this chapter, we highlight the major phases of the research process in which research-design equivalence needs to be established to ensure that the results from one's study can be attributed to true group differences: problem formulation, research design, sampling, measurement selection, and data collection (except for the data analysis phase, which will be discussed in Chapter 3). Additionally, a detailed table (see Table 2.1) is provided that presents an overview of what types of equivalence should be established at each phase, and the strategies that need to be used to do so.

PROBLEM FORMULATION

During the problem formulation stage of the research process, a central concern is to ensure that the construct under investigation is conceptualized similarly across groups. To do this, construct equivalence must be established. *Construct equivalence* refers to whether the same concept is

Table 2.1 Establishing Equivalence at Different Stages of the Research Process

Stage	*Concern*	*Specific Issues*	*Potential Remedies*
Problem Formulation	**Construct equivalence: Same factor structures obtained across groups**	**Conceptual/Configural**—whether different populations conceptualize the construct that the measure is assessing in the same manner	Conducting a multi-group confirmatory factor analysis.
		Functional—whether the construct functions in a similar manner across the groups (i.e., relates to other concepts in a similar manner or as theoretically expected)	Conducting a multi-group analysis, where the structural paths are unconstrained across both groups; and compare it with the fully constrained based model.
		Scalar—exists when the scores on the measure assessing the construct are similar in strength across the groups	See Chapter 3.
Research Design	**Threats to causal inferences**	**Statistical conclusion validity**—threats include inadequate statistical testing, inadequate testing for structural equivalence, and inadequate testing for scalar equivalence	Procedures such as structural equation modeling, differential item functioning, and item response theory analysis are statistical methods that can be used to address the threats to statistical conclusion validity.
		Internal validity-selection and differences in response styles—non-equivalent construct definition, non-equivalent operational definition, differential familiarity with research materials and settings, reactivity to the research setting, and experimenter expectancies	One way to minimize the effects of selection, a threat to internal validity, on one's results is to match the groups on a variable that is known to be associated with the dependent variable.

		Construct validity—non-equivalent construct definition, non-equivalent operational definition, differential familiarity with research materials and settings, reactivity to the research setting, and experimenter's expectations	Threats to construct validity, such as non-equivalent construct definition, can be minimized by carefully reviewing the literature to determine if the measure one plans to use is both valid and reliable for the groups involved in one's study.
Sampling	**Confounding factors**		Matching participants on confounding factors; propensity score matching. Statistically control for other factors representing rival explanations. Collect data on potential confounding variables and control for them statistically using hierarchial multiple regression analysis of covariance.
	Within-group diversity: Subpopulations of a targeted population	**Composition effect**—certain individuals may have a higher probability of getting into the sample than others	Select both samples in a similar manner. Conduct a latent class analysis to identify subpopulations from a heterogeneous sample.
Selection of Measures	**Construct equivalence: Same factor structures obtained across groups**	**Conceptual/Configural**—whether different populations conceptualize the construct that the measure is assessing in the same manner	It is important that the measures be normed, valid, and reliable for the groups being surveyed.
	Acquiescence response style: Effects of response styles can increase the risk of making a Type I or Type II error	**Type I error** produces results that indicate that there are "true" differences between the groups when no such differences exist	Attention to ambiguity of the question and read-ability of the question; both are associated with an acquiescence response style. Attention to statistical methods.

(Continued)

Table 2.1 (*Continued*)

Stage	*Concern*	*Specific Issues*	*Potential Remedies*
		Type II error indicates that no differences between the groups exist when really there are true differences between the groups	
Data Collection Only one mode of data collection should be used to collect the data from the groups involved in one's study, to minimize the effects of response styles on one's results.	**Procedural equivalence:** Ensuring that there is consistency across groups in the way the surveys are administered, the timing of survey administration, the conditions under which the surveys are administered, and the mode of data collection (Schaffer & Riordan, 2003)	**Surveys**	When conducting research with diverse groups, one should use the same means of distributing the surveys. Develop a similar cover letter and instructions for completing all surveys.
		Interviews	Procedural equivalence still must be established for interviews through standard procedure for training all the interviewers. Once the interviewers have been trained, periodic retraining on the administration of the surveys should also been done, to ensure that the interviewers are administrating the survey properly.
		Timing	Executing the survey during the same time frame will reduce the probability that external stimuli are affecting one's results.
Data Analysis Not assessing for these three types of equivalence (among others) can inflate the risk of Type I and Type II error	**Conceptual/Configural equivalence**	Conducting a multi-group confirmatory factor analysis Conducting Exploratory (EFA) and Confirmatory Factor Analysis (CFA)	
	Metric equivalence	Conducting a multi-group confirmatory factor analysis, where you impose equality on the factor loadings across the groups and fit the model to the data for each group simultaneously	
	Scalar equivalence	Conducting a multi-group confirmatory factor analysis, where the intercepts are unconstrained across both groups, and comparing it with the fully constrained based model	

being measured across groups. Construct equivalence exists when the same factor structures are obtained across groups. Establishing construct equivalence assures that both groups conceptualized the construct under investigation the same way, and that the research findings are due to the construct of interest and not to other sources, such as measurement error. Without construct equivalence, there is no basis for group comparisons (van de Vijver & Leung, 1997). It is important to establish several types of construct equivalence: conceptual, functional, and scalar. *Conceptual/Configural equivalence* refers to whether different populations conceptualize a construct that the measure is assessing in the same manner. *Functional equivalence* refers to whether a construct functions similarly across groups (i.e., it relates to other concepts similarly or as theoretically expected; Adamsons & Buehler, 2007; Knight, Virdin, Ocampo, & Roosa, 1994). For example, functional equivalence is established if the correlation between depression and self-worth is the same across groups (Knight et al., 1994). *Scalar equivalence* exists when the scores on the measure assessing the construct are similar in strength across groups. For example, scalar equivalence exists when the cutoff score indicating clinical depression is the same for each group (Crockett, Randall, Shen, Russell, & Driscoll, 2005).

RESEARCH DESIGN

Most studies conducted with diverse groups employ a non-experimental, comparative research design. A research design is non-experimental when researchers are not concerned with demonstrating causality. A research design is comparative when two or more naturally occurring groups are compared. Because non-experimental research designs do not demonstrate causality and only show that a relationship exists, it is important for researchers using these types of designs to rule out alternative explanations for the obtained findings.

Leung and van de Vijver (2008) identified several threats to causal inferences in cross-cultural research that are important to consider when conducting research with diverse groups. These threats are *statistical-conclusion validity*, *internal validity*, and *construct validity*. For Leung and van de Vijver, threats to statistical-conclusion validity include inadequate statistical testing in general and inadequate testing for structural and scalar equivalence in particular. Differential item functioning, item response theory analysis, and structural equation modeling (SEM) are statistical methods that can be used to address the threats to statistical-conclusion validity (for a more detailed discussion of these procedures, please refer to Jöreskog, 1971; Muthén, 1989; Reise & Haviland, 2005; Teresi, 2006). According to Leung and van de Vijver, threats to internal validity include selection and differences in response styles. *Selection* is used to describe a situation in which the groups differ prior to being selected to participate in the study, and these differences, rather than the variable of interest, account for the results. One way to minimize the effects of selection on the results is to match the groups on a variable that is known to be associated with the dependent variable. By doing this, the researcher maximizes the chances of obtaining true group differences. Threats to construct validity include nonequivalent construct definitions, nonequivalent operational definitions, differential familiarity with research materials and settings, reactions to the research setting, and experimenter's expectations (Leung & van de Vijver, 2008). Threats to construct validity can be minimized by carefully reviewing the literature to determine whether a measure is both valid and reliable for the groups involved in the study. All of the threats need to be addressed, or the conclusions are likely to be erroneous.

SAMPLING EQUIVALENCE

Valid conclusions from studies comparing diverse groups necessitate samples that differ only in terms of (for instance) gender or ethnicity. To ensure that valid conclusions are drawn, it is important that rival alternative explanations be minimized. Confounding variables are one cause of competing explanations, and as such they represent potential threats to a study's validity. One way of lessening the effects of potential confounding factors is to match participants on those very factors so that the study groups are comparable. Confounding factors are usually demographic characteristics of the sample population that studies have found to impact the outcome variable. Researchers should be aware that matching is not without its limitations. For example, matching may result in statistical regression toward the mean; for instance, people in Group A might have extremely high scores on the outcome variable, whereas people in Group B might have extremely low scores on the same variable. In this case, people in Group A would seem to have better scores than they really have, and people in Group B would seem to have worse scores than they really have. In this example, the difference between the groups is not due to the group being selected on the basis of extremely high or extremely low scores on the outcome variable; rather, the scores for the confounding variable did not overlap, and the researcher did not detect this problem until after the samples had been matched. Furthermore, matched samples are not representative of the larger population, and therefore the results cannot be generalized beyond the study sample.

Although matching can be used as a method to establish sampling equivalence, researchers still need to statistically control for other factors representing rival explanations. To do this, researchers should gather information about other potential confounding variables that were not controlled for via matching. In instances where matching cannot be used as a way to obtain comparable samples, the researcher should still collect data on the potential confounding variables and control for them statistically, using hierarchical multiple regression analysis of covariance.

As an alternative to matching, propensity score matching can be used to establish sampling equivalence. Propensity score matching minimizes selection effects, allowing for less biased comparisons between groups on the outcome variable (Guo, Barth, & Gibbons, 2006). Propensity score matching employs a predicted-probability membership based on identified (observed) confounding variables, such as demographic characteristics of the sample, and is usually based on logistic regression to create a counterfactual group (Baser, 2006). Propensity scores can be used to match the sample. According to Berg, Johnson, and Fleeger (2003), "matching on a propensity score is a way of matching on many variables indirectly, instead of matching directly on many variables, which becomes increasingly difficult with more variables" (p. 739). In essence, propensity scores comprise several confounding variables used to match the sample. Selecting the appropriate confounding variables to include or exclude in the propensity score matching is a critical initial step, because omitting important confounding variables produces inaccurate propensity scores (Baser, 2006; Polsky et al., 2009). Variable selection should be based on empirical studies that demonstrate the interrelationships among the variables of interest (Smith & Todd, 2005). Several different matching techniques can be employed with propensity score matching, and each can produce different results (for more information on the different techniques, see Baser, 2006; and Smith & Todd, 2005).

Another potential threat to the validity of studies comparing diverse groups is within-group diversity. *Within-group diversity* is defined as the subpopulations of a target population. For example, the U.S. Hispanic population includes Mexican Americans, Cuban Americans, Dominicans, Puerto Ricans, and many other subpopulations of Hispanics. These subpopulations differ in their use of language, their acculturation, and their immigration status in the United States (Knight et al., 2009). This is just one example of why it is important to clearly define the population target.

By not acknowledging within-group diversity and therefore failing to disaggregate variables that influence the outcome of interest, researchers may overestimate the homogeneity of the population under investigation (Alfredo, Rueda, Salazar, & Higareda, 2005). Research that overlooks within-group diversity may lead to misguided interventions and policies intended for these populations (Knight et al., 2009). Therefore, when conducting research with diverse racial and ethnic populations, researchers must make a concerted effort to account for within-group diversity. Researchers have attempted to account for within-group diversity when conducting research with Hispanics, for example, by assessing their acculturation, generation status, and language spoken in the home (Bernal, & Sharron-del-Rio, 2001; Knight et al., 2009). Another way to account for within-group diversity is to establish homogeneity by using latent-class analysis. Latent-class analysis is used to identify subpopulations from a heterogeneous sample. Readers who are interested in learning more about latent class analysis can refer to Clogg (1995), Neely-Barnes (2010), and Rosato and Baer (2012).

When conducting research with diverse groups, it is important that both samples be selected in a similar manner, to avoid introducing bias with the sampling method. In looking at the patterns of results across four studies that used different sampling methods, Bowen, Bradford, and Powers (2007) found that sexual-minority women who were selected by probability sampling methods had lower rates of mammography screening than sexual-minority women who were selected by non-probability sampling methods. Furthermore, compared to the sexual-minority women selected via non-probability sampling methods, the probability sample had a lower percentage of sexual-minority women who had a college education or were in a relationship.

The use of different sampling methods can also produce a composition effect, where certain individuals have a higher probability of getting into the sample than others, especially when stratified sampling is used to select one group of participants, and convenience sampling is used to select another. By using stratified sampling, the researcher is intentionally trying to select participants for the study based on certain characteristics, such as income status or educational attainment. Furthermore, the use of different sampling methods can lead to unequal sampling errors across groups and reduce construct validity, thereby threatening the validity of the findings (Kumar, 2000).

MEASUREMENT SELECTION

When selecting measures, researchers should keep conceptual equivalence in mind. Recall that conceptual equivalence refers to whether different populations conceptualize the construct that the measure is assessing in the same manner. If the construct is conceptualized differently for each group, then the measure may not accurately capture the construct for each group (Crockett et al., 2005). The use of an inaccurate measure could then lead to an underestimation or overestimation of prevalence rates for the groups involved in the study (Crockett et al., 2005).

In conducting research with diverse groups, it is also important that the measures be normed, valid, and reliable for the groups being surveyed. It has been found that measures that were valid and reliable for males were not necessarily valid and reliable for females (Ibrahim, Scott, Cole, Shannon, & Eyles, 2001; Orhede & Kreiner, 2000). Similarly, it has been demonstrated that measures that were valid and reliable for Whites were not always valid and reliable for African Americans (Kingery, Ginsberg, & Burstein, 2009).

In selecting a measure, researchers should make sure that the measure itself does not produce response styles. Questions that are worded ambiguously (Ray, 1983) or are difficult to read (Stricker, 1963) tend to produce acquiescent responses. An *acquiescent response style* occurs when one agrees with the survey item regardless of the content. Response styles may produce bias in the true score by inflating or deflating the observed score, and may produce bias in the interrelationships among the variables of interest by inflating or deflating the correlations among the variables (Baumgartner & Steenkamp, 2001). In other words, the effects of response styles can increase the risk of making a Type I or Type II error. *Type I error* produces results that indicate that there are true differences between study groups when no such differences exit. *Type II error* indicates that no differences between study groups exist when really there are true differences between the groups.

It is particularly important to consider response styles when conducting research with diverse groups because studies have shown that Hispanics and African Americans in the United States show higher tendencies toward extreme response styles (Johnson, Shavitt, & Holbrook, 2011). Response styles have implications for measurement equivalence. Acquiescence can affect scalar equivalence, whereas extreme response styles can affect both metric and scalar equivalence (Kankaras & Moors, 2011). If the effects of response styles go undetected, then the results will not be reflective of true group differences. Therefore, researchers need to test for the effects of response styles on measurement equivalence. An in-depth discussion of the various statistical methods that can be used to detect response styles is beyond the scope of this book. For those interested in learning more about these methods, please refer to Baumgartner and Steenkamp (2001), Cheung and Rensvold (2000), Kankaras and Moors, (2011), Moors (2003, 2004), and van Herk et al. (2004).

Another issue that researchers should consider when selecting a measure is negatively worded items. The effect of negative wording on participant responses is known as a *method effect*. Method effects are a source of measurement error that has implications for the interpretation of scores within and between groups (Marsh, Scalas, & Nagengast, 2010). According to Barnette (2000), "negatively worded items are those [items] phrased in the opposite semantic direction from the majority of the items on the measure" (p. 361). Negatively worded items usually contain the word *not*. Generally, negatively worded items on a survey are recommended as a way of reducing an acquiescent response style (Cronbach, 1950). Research, however, has demonstrated that negatively worded items can reduce the reliability of a measure (Barnette, 2000), produce spurious factors consisting predominately of the negatively worded items (Ibrahim, 2001), and result in a unidimensional measure becoming a multidimensional measure (Chen, Rendina-Gobioff, & Dedrick, 2010). Given the issues noted above, negatively worded items on surveys with diverse groups may have serious implications for establishing construct, conceptual, functional, and scalar equivalence. For example, in examining the scalar equivalence of a Chinese self-esteem scale for third- and sixth-graders, Chen et al. (2010) found that both groups had the same level of self-esteem but had different average item scores. The authors concluded that the differences between the means for these two groups should be interpreted with caution, as they are not reflective of true group differences but are due to differences in how the third-graders were influenced by the negatively worded items.

The preferred statistical method for examining the effects of negative wording effects is a CFA (DiStefano & Motl, 2006), because it overcomes the shortcomings of an exploratory factory analysis (Chen et al., 2010). For more information about conducting a CFA to examine negative-wording effects, please refer to Chen et al. (2010) and Supple and Plunkett (2011).

DATA COLLECTION

After designing the study, developing the survey, and determining the sample strategy, researchers must focus on establishing procedural equivalence. *Procedural equivalence* refers to ensuring consistency across groups in the way the surveys are administered, the timing of the surveys' administration, the conditions under which the surveys are administered, and the mode of data collection (Schaffer & Riordan, 2003).

Ensuring that the survey is administered in the same way across groups is an important step in establishing procedural equivalence. Differences in the administration of the survey can lead to different response rates and results that are erroneously attributed to group differences (Steinmetz, Schwens, Wehner, & Kabst, 2011). Steinmetz et al. (2011) recommended that, when conducting a cross-cultural study, researchers should use the same means of distributing the survey to all potential participants, and develop a similar cover letter and instructions for completing the surveys. We recommend that these same procedures be followed when conducting research with diverse groups as a way to establish procedural equivalence. A change in the environmental conditions during the data collection process may bias the results (Schaffer & Riordan, 2003); therefore, researchers need to ensure that the surveys are administered under the same conditions for all groups in their studies.

When conducting interviews with participants in a study (rather than administering self-report surveys), procedural equivalence still must be established. One way to establish procedural equivalence is to have a standard procedure for training all the interviewers. Once the interviewers have been trained, periodic retraining on the administration of the surveys should also be done to ensure that the interviewers are administering the survey properly.

The timing of a survey's administration can affect the validity of the conclusions. It is important to administer the survey for both groups during the same time frame; doing this will reduce the effects of external stimuli on the results. For example, if one were studying the effects of past traumatic events on adolescent mental health, it would be important that all of the participants be surveyed during the same period so that the results are not affected by a recent traumatic event, such as a school shooting that occurred the day of the survey's administration.

The mode of data collection needs to be the same across groups, because different modes of data collection are associated with different response styles. For example, in comparing the use of telephone and household interviews, Jordan, Marcus, and Reeder (1980) found that telephone interviews resulted in higher acquiescence and extreme response styles. In comparing the use of self-administered surveys, online surveys, and telephone surveys, Weijters, Schillewaert, and Geuens (2008) found that telephone surveys resulted in lower midpoint responses but higher levels of acquiescence and extreme responses. Based on the findings of their study, they cautioned against using telephone surveys in conjunction with self-administered or online surveys. Meanwhile, in a meta-analytic study, it was found that computer-assisted telephone interviews resulted in extreme positive responses when compared to other modes of data collection, such as face-to-face and mailed surveys (Ye, Fulton, & Tourangeau, 2011).

Recall that the effects of response styles can increase the risk of making a Type I or Type II error. Given the issues describe above, we recommend that only one mode of data collection be used to collect data from groups in one's study to minimize the effects of response styles on the results.

Moreover, the mode of data collection can affect the quality of the data in several ways. One way is by influencing the way people respond (i.e., their response set). Research has demonstrated that self-administered surveys can increase the likelihood of people responding to sensitive questions (e.g., alcohol consumption and illicit drug use), whereas face-to-face interviews can decrease the likelihood of people responding to sensitive questions (Tourangeau & Smith, 1996). The mode of data collection can also influence who responds to the questions (*compositional effect*; Elliott et al., 2009). A compositional effect occurs, for example, when a survey is administered online and by telephone, and the researcher finds out that younger people were more likely to respond to the online survey than to the telephone survey. Yet another way in which the mode of data collection can affect the quality of the data is by producing a response-choice effect. Research has demonstrated that when participants complete a self-administered survey, they tend to select the first response category (primacy effect), whereas people who are surveyed via telephone tend to select the last response category (recency effect; Bowling, 2005). For more information about other ways in which the mode of data collection affects the quality of the data, please refer to Bowling (2005).

DATA ANALYSIS

We have already described what one needs to do at the various phases of the research design (i.e., problem formulation, research design, sampling, and data collection) to ensure that equivalence is established in studies with diverse groups. Similarly, it is also essential to establish equivalence in the data analysis phase. The next chapter includes an in-depth discussion of the seven types of equivalence that need to be established when conducting research with diverse groups during the data analysis phase. Additionally, we demonstrate how MG-CFA can be used to assess these types of equivalence.

SUMMARY

The educational, economic, health, and social disparities between diverse groups will continue to be a driving force for conducting research with these groups so that knowledge can be generated to understand why these disparities exist. Much of the research to date has been conducted where researchers have not paid adequate attention to ensuring that equivalence has been established at all phases of the research process. The lack of this attention may have inadvertently resulted in methodological flaws and erroneous inferences drawn from these studies. Ideally, if greater attention is paid to establishing equivalence at each phase, then it is more likely that the results can be attributed to true group differences.

Research-design equivalence refers to the processes and procedures that ensure accurate representation of the phenomenon under investigation across diverse groups. In this chapter, we highlighted the phases of the research process to which attention must be given to achieve research-design equivalence: problem formulation, research design, sampling, measurement selection, and data collection (except for the data analysis phase, which will be discussed in Chapter 3). Methodological issues that may result in non-equivalence across groups were discussed, and strategies that can be used to enhance equivalence across groups were presented. In addition, we discussed the importance of ruling out alternative explanations for one's findings associated with using a non-experimental, comparative research design, which is most often used when conducting research with diverse groups.

We hope that this chapter provided an excellent overview of the processes and procedures needed to achieve research-design equivalence across diverse groups. It is important that these processes and procedures be implemented prior to the data analysis phase, as the results of the analysis could still indicate that measurement equivalence exists, despite flaws in any phase of the research process. We will provide a more detailed discussion about the various types of measurement equivalence and the statistical approaches used to establish them in Chapter 3.

It is equally important that research-design equivalence be achieved at the data analysis phase. In Chapter 3, we detail what processes and procedures need to be implemented during the data analysis phase to achieve research-design equivalence. These processes and procedures are used to establishment measurement equivalence—a necessary step that should proceed any multivariate analyses that are used to test the hypotheses for one's study. Chapter 3 will expose readers to seven types of measurement equivalence, which at a minimum need to be established when conducting research with diverse groups: configural, metric, scalar, error (i.e., covariance), factor variances, factor covariances, and factor means. The rationale for establishing each type of measurement equivalence is presented, along with the statistical strategy used to assess that particular measurement equivalence, within a MG-CFA framework. Specifically, we focus on establishing measurement equivalence across groups when the groups have been identified based on directly observable characteristics (e.g., gender, ethnicity) or manifest variables. To illustrate the concepts discussed in Chapter 3, a case example is presented. Mplus is used to analyze the data for the case illustration. The Mplus syntax for the MG-CFA analyses are provided. A write-up of the case example results are presented as if they were being reported in a publication. Additionally, a hypothetical case example is presented in Chapter 4 to further illustrate the analytical procedures used to establish equivalence. This case example also illustrates how descriptive statistics (means, standard deviations, skewness and kurtosis) and can be examined to determine initially whether measurement equivalence has been established across the groups.

3

Multi-Group Confirmatory Factor Analysis to Establish Measurement and Structural Equivalence

OVERVIEW

In the previous chapter, we described the major phases of the research process where equivalence needs to be established. In this chapter, we will demonstrate how to apply an MG-CFA to establish the equivalence of measurement instruments used to assess the construct of interest when conducting research with diverse groups. The MG-CFA analyses described in this chapter use a covariance structure only (COV) and mean and covariance structures (MACS). When an MG-CFA is based only on a COV, the configural model, factor loadings, residual covariance, factor variances, and factor covariances can be tested for equivalence. Incorporating the mean and covariance structures allows researchers to test for equivalence of the intercepts and latent means. Testing for the equivalence of the intercepts is necessary to determine whether a comparison of the observed means can be done. Meanwhile, testing for the equivalence of the latent means can be used to determine whether the two groups differ at the level of the construct's unobserved mean.

This chapter focuses on measurement equivalence across groups when the means for identifying the groups are based on directly observable characteristics (e.g., gender, ethnicity) of the participants or manifest variables. Although measurement equivalence has also been applied to situations when the timing of the administration of the measure is the defining feature of the grouping (longitudinal measurement equivalence), we do not discuss this type of equivalence. Readers interested in learning about establishing measurement equivalence in longitudinal studies can refer to Willoughby, Wirth, and Blair (2012).

Readers should also note that, within a CFA framework, single-group multiple-indicator, multiple-cause (MIMIC) structural equation models can be used to investigate measurement equivalence (Ackerman, 1992; Fleishman, 2004; Millsap & Everson, 1993). The MIMIC approach offers several advances over the MG-CFA approach. For example, smaller sample sizes are needed, and a MIMIC model does not require that the potential invariant variable be categorical. Unlike MG-CFA models, MIMIC models do not give researchers the ability to investigate the potential non-equivalence of the factor loadings, observed residual variances or covariances, intercepts, factor variances, covariances, or means. Because MIMIC models can only be used by researchers to study whether the means of the factors vary as a function of the potential invariant factor, MG-CFAs offer a more comprehensive approach to the study of measurement equivalence than MIMIC.

The next section discusses the meaning of measurement equivalence. We begin by providing a definition of measurement equivalence based on a latent-variable framework, and then discuss the steps used to establish measurement equivalence using an MG-CFA, based on both a COV and MACS.

MEASUREMENT EQUIVALENCE DEFINED

Measurement theorists provide definitions of measurement equivalence that focus on the consistency of the relationship between latent variables and observed variables among groups of individuals completing a measurement instrument (Meredith, 1993; Meredith & Teresi, 2006). A measurement instrument is said to be *invariant* between groups when the probability of an individual receiving a particular observed score is not dependent on his or her group membership but is dependent on his or her true score (Wu & Zumbo, 2007). In other words, individuals in different groups (e.g., different gender or cultural groups) who have the same true score or level of the latent variable being measured will have the same observed score if the measure is invariant between the groups (Meredith, 1993). This definition applies when researchers are concerned about the invariance of a particular question on a measure and the measure as a whole. In sum, when using an MG-CFA framework a measure has between-group equivalency when the mathematical function that relates latent variables to the manifest variables (indicators) is the same in each of the groups (Borsboom, 2006).

In addition to the statistical definition of measurement equivalence, Meade and Bauer (2007) provided a conceptual definition of measurement equivalence: "Measurement equivalence can be considered the degree to which measurements conducted under different conditions yield equivalent measures of the same attributes" (p.611). According to Meade and Bauer, these different conditions can be characteristics of the participants in the study, such as their age, gender, or ethnicity. This conceptual definition of measurement equivalence focuses our attention on the equivalency of the meaning of the measure. This leads to the question of what evidence is needed to determine the between-group equivalency of a measure. If researchers find evidence that the psychometric properties are equivalent across groups, then the measurement equivalence–argument is supported (Ashton & Paunonen, 1998).

Lubke, Dolan, Kelderman, and Mellenbergh (2003) noted that a measure can be unbiased when comparing two particular ethnic groups but biased when comparing gender groups within those ethnic groups. Therefore, it is necessary to test for measurement equivalence across all dimensions of the study groups when comparisons are made (Meredith, 1993). Additionally, once there is evidence that a particular group dimension is not associated with bias, researchers can assume that other characteristics that are highly associated with that dimension will also be unbiased (Lubke et al., 2003).

This only holds true when strict measurement invariance is found, and does not hold true for lesser forms of measurement invariance (Lubke et al., 2003).

Equivalency of a measure's properties is a matter of degree. Equivalency of the measure's properties does not require the complete absence of any non-equivalence. Group-specific elements of a measurement model's structure will often occur, resulting in partial non-equivalency measure (Ashton & Paunonen, 1998). For example, it is possible for patterns of fixed and free parameters in the measurement model to differ slightly between groups (i.e., one item can cross-load on factors for one group but not for the other group) with no other evidence of measurement non-equivalence between the groups. Additionally, the impact of measurement equivalence on a measure's utility is related to both the location of the equivalence and the context of the measure's use (Borsboom, 2006). The following section presents an overview of the various analytical procedures used to establish the measurement equivalence of an instrument.

OVERVIEW OF MULTI-GROUP CONFIRMATORY FACTOR ANALYSIS

In his seminal work, Jöreskog (1971, 1993) designed the process of using an MG-CFA to evaluate measurement equivalence. The equivalence of the factor structures between the two groups is tested using a series of hierarchically nested MG-CFA models (Jöreskog, 1971). The models are ordered with increasing numbers of equivalency constraints imposed on parameters between the two groups. The equivalence of parameters between the two groups is tested to determine the degree and location of between-group measurement equivalence.

Within a CFA framework, there are two types of variables—manifest and latent. The manifest variables are directly observable from items on the measure, while the latent variables are estimated from the manifest variables (Klem, 2000); therefore, there is a causal relationship between the latent and manifest variables. The variation in the manifest variable (y_{i1}) is a function of the latent variable, specific factor variable, and error of measurement. In many cases, the specific factor variable and error of measurement are lumped together and referred to as *uniqueness*. When the analysis of the structural equation model or factor analytic model is based on a COV, the mathematical expression that represents this factor model for an individual is represented by equation 1 below:

$$y_{il} = \lambda_i \eta_i + \lambda_i \eta_i + \ldots + \lambda_{jm} \eta_{mi} + \varepsilon_{ji} \qquad (1)$$

where y_{il} identifies each observed variable, j refers to the jth observed variable ($j = 1, \ldots p$), and i denotes the ith individual ($i = 1 \ldots N$).

In matrix form, $y_i = v + \Lambda\eta_i + \varepsilon_i$ where:

v is the vector of intercepts v_j

Λ is the matrix of factor loading λ_{jk}

ψ is the matrix of factor variances/covariances

Θ is the matrix of residual variance/covariances

The population covariance matrix of observed variables Σ,

$$\Sigma = \Lambda\psi\Lambda' + \Theta$$

The equivalence between groups for each parameters responsible for variation in the observed variable (y), the factor-loadings matrix ($\wedge'$), the factor variances–covariances matrix (ψ), and the residual variances–covariances matrix (Θ) can be tested. When the analysis is based on the covariance structure, the sample mean and sample covariances are not present in the model. The absence of the sample mean and covariances from the model prevents testing of the intercepts and the equivalence of the latent mean structure (Byrne, 2012). The incorporation of the means into the analysis allows for the testing of "stronger" forms of measurement equivalence (Meredith, 1993).

Widaman and Reise (1997) demonstrated that, to include the mean structure in the analysis, the scores for the indicators are left in their raw form. Therefore, the equation for predicting the indicator becomes:

$$Y_{ij} = \tau_j + \lambda_{i1}(\alpha_1 + \eta_{1i}) + \lambda_{i2}(\alpha_2 + \eta_{2i}) + \cdots + \lambda_{m2}(\alpha_m + \eta_{mi}) + \varepsilon_{ji} \quad (2)$$

In the equation, τ_j is the intercept term predicting the observed manifest variable (Y_{ij}). The other predictors include the latent variable (η_{mi}), the mean of the latent variable (α_m), and the factor loading (or regression coefficient, λ). When carrying out a measurement-equivalence analysis using a COV, equivalence can be tested for factor loadings, factor variance and covariance, and residual variance and covariance. Adding the latent mean structure to the analysis allows the researcher to test the factor intercepts and latent means.

When investigating the measurement equivalence of an instrument across groups, it is possible to test the equivalence of all measurement and structural parameters common to the groups (Meredith, 1993). Parsimony needs to be a driving force during this process. There is growing consensus that when researchers are concerned with establishing the construct validity of a measure, the analysis should focus on the equivalence of factor loadings, intercepts, variance, and covariance (Byrne, 2012). When the goal is to compare group means, testing the measurement equivalence of the latent means is appropriate (Ployhart & Oswald, 2004).

TESTING MEASUREMENT EQUIVALENCE ACROSS GROUPS

When conducting research with diverse groups, an underlying assumption is that the researcher is measuring the same construct across all groups in the study. Oftentimes this assumption is taken at face value and is not tested; however, not testing the assumption can result in erroneous conclusions (Milfont & Fischer, 2010). Testing for measurement equivalence is a necessary and important step in conducting research with diverse groups. The following discussion presents the seven types of measurement equivalence that are important to establish when conducting research with diverse groups (Milfont & Fischer, 2010). These seven types of equivalence can be classified into two categories: measurement and structural. *Measurement equivalence* focuses on establishing equivalence by examining how items function across groups. Measurement equivalence can be assessed by examining factor loadings, item intercepts, and error variances across groups (these are considered to be the *observed variables*; Milfont & Fischer, 2010). *Structural equivalence*, on the other hand, refers to establishing the theoretical structure of the measure across groups (Byrne & Watkins, 2003). Structural equivalence can be assessed by examining factor variance, factor covariance, and factor means across groups (these are considered to be *unobserved* or *latent variables*).

Evaluating an instrument's measurement equivalence requires estimating several nested models (see Table 3.1). *Nested models* are models in which the parameters are contained within one another in a specific hierarchical order. For example, in a regression analysis where gender and age are predictors, a model that contained only gender as a predictor would be said to be nested within the model that includes both gender and age as predictors. Model 0 (Baseline CFA) consists of separate group analyses to evaluate the descriptive statistics and factor structures of the measure for each group. Model 1 evaluates configural equivalence by estimating an MG-CFA model that does not contain any between-group equivalency constraints. Model 2 evaluates weak metric equivalence and involves estimating an MG-CFA that imposes equivalency constraints on the factor loadings. Model 3 tests for strong metric (scalar) equivalence and involves estimating an MG-CFA that imposes equivalency constraints on factor intercepts. Model 4 tests for strict metric equivalence (error variances and covariances) and involves estimating an MG-CFA that imposes equivalency of the regression residual variances and covariances. Model 5 evaluates the equivalence of latent means and involves estimating an MG-CFA that imposes equivalency constraints on the factor (latent) means. Model 6 tests for equivalence of the factor variance and involves estimating an MG-CFA that imposes equivalency constraints on the factor (latent) variances. Finally, Model 7 evaluates the equivalence of factor covariances and involves estimating an MG-CFA that imposes equivalency constraints on the factor (latent) covariances.

Table 3.1 Types of Measurement Equivalence

Necessary	*Model*	*Question*	*Type*	*Parameter Equivalency Test*
YES	0	Can a CFA model be established for each group separately?	Base CFA	Separate group CFA/EFA
	1	Do the indicators making up a particular measure operate equivalently across different groups?	Configural equivalence	The number of factor and loading pattern examined (no equivalency constraints imposed).
	2	Is the construct validity of the measure equivalent across groups?	Weak metric equivalency	Assuming partial or full configural equivalency, the regression coefficient (factor loadings) are constrained.
	3	Can an unbiased comparison between groups be made using the observed means for each group?	Strong metric (scalar) equivalency	The regression intercept term, needed if mean differences will be examined.
NO	4	Does the measure have the same reliability for each group?	Strict metric (error variance/covariance) equivalency	The regression residual variance. This is only a test of the reliabilities when the factor variances are equal.
	5	Are there differences between the groups on the latent means?	Equivalency of latent means	If scalar equivalency exists, the means of the common factors are constrained.
	6	Is there equal heterogeneity of the latent variables in the groups?	Equivalency of factor variance	The variances of the common factors.
	7	Is there equivalence in the association between factors across groups?	Equivalency of factor covariances	The covariances among the common factors.

Note: All analyses involved the use of Multiple Group Confirmatory Factor Analysis, with the exception of Model 0.

Model 0: Separate Group Analysis

The separate group analysis involves both an examination of the distributional properties of a measure and the factor structure (see Figure 3.1). Researchers must examine the similarity of the items' distributions, observed scale means, variances, skewness, and kurtosis. Differences in skewness between groups can reflect particular patterns of responses associated with response styles. Groups that demonstrate different levels of acquiescence in their responses will have different levels of skewness (Byrne & Campbell, 1999). Although it has been recommended that a measure's reliability (i.e., coefficient alpha; Nimon & Reio, 2011) should be evaluated as a part of the separate group analysis, the structure of the measure should be evaluated prior to evaluating the reliability of the measure in order to determine the appropriate method for estimating the reliability of the scale (Yanyun & Green, 2011). A factor analysis should be run to determine the number of dimensions present in the measure. When there is evidence that the measurement model may be multidimensional, a multidimensional or composite reliability coefficient can be used (Byrne & Campbell, 1999; Widhiarso, 2010). Scott Colwell at the University of Guelph has developed a web-based calculator (http://wwwi.uoguelph.ca/~scolwell/cr.html) for estimating the composite reliability of a measure using the factor loadings and error variances. Additionally, Mplus

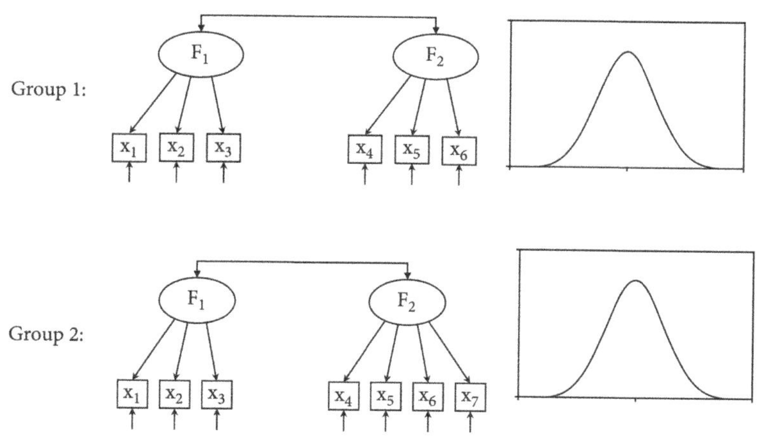

Figure 3.1 Model 0: Separate Group Base Models

syntax is available to estimate the composite reliability. (See the Mplus website—http://statmodel.com/download/usersguide/Mplus%20 user%20guide%20Ver_7_r6_web.pdf).

Following the descriptive analyses, an exploratory factor analysis (EFA) and CFA analysis need to be conducted for each group separately to determine the best-fitting CFA model. A single-group CFA analysis is used to determine the areas of the measurement model that are common between the two groups. The goal of this analysis is to determine the best-fitting measurement model for each group. The model-evaluation process should include both statistical and theoretical criteria. Statistical criteria will involve the examination of fit indices, for example, Confirmatory Fit Index (CFI). Root-Mean-Squared Error of Approximation (RMSEA), Standardized Root Mean Squared Residual (SRMR). Additionally, using the Modification Index (MI) to make post hoc model modifications and conduct sensitivity analyses is part of the process for identifying the baseline model (Hayduk & Glaser, 2000; Millsap, 2007). In addition, following the recommendations of Bollen (1989) and Gonzalez and Griffin (2001), the overall fit indices of each parameter should be examined for statistical significance and evaluated against the theoretical framework used in the study. The evaluation of fit indices to determine the best-fitting model must take into account the sample size, the number and magnitude of the factor loadings, and the specific area in which misspecification has occurred.

Although the analyses of the baseline measurement model for each group aim to establish commonality in the patterns of free and fixed parameters between the groups, specificity of measurement instruments can lead to a failure to establish full equivalence on the patterns of the factor loadings across groups (Byrne, Shavelson, & Muthén, 1989). When this occurs, the researcher can proceed with further measurement-equivalence analyses, although only partial-configural equivalence can be evaluated.

A meaningful level of commonality in the pattern of fixed and free parameters between the two models is needed to justify further analyses of measurement equivalence (Byrne, 2012; Vandenberg & Lance, 2000). Researchers should be aware that the exact amount of commonality needed to support further testing of equivalence is not clear (Yoon & Millsap, 2007). As with the criteria used to evaluate other CFA models, the researcher needs to consider theory and empirical information on the validity of the measurement model to determine the amount of commonality that is needed to proceed with further analyses of measurement equivalence.

After the preliminary descriptive analysis of each group by itself, a series of MG-CFA models needs to be conducted to test the equivalence of the various parameters that make up both measurement and structural aspects of the given measure model (Byrne et al., 1989; Byrne & van de Vijver, 2010; Milfont & Fischer, 2010). The models are organized in a hierarchical order to test the equivalence of specific model parameters across groups. Moreover, the models are ordered with a decreasing number of freely estimated parameters between the groups. The fit of these increasingly restrictive models is used to evaluate the extent of the measurement equivalence (Vandenberg & Lance, 2000).

Model 1: Configural Equivalence

This model investigates whether the indicators making up a particular measure operate equivalently across different groups and whether initial evidence exists of the face validity of the measure across groups. To address this issue, researchers must test for configural equivalence. Figure 3.2 depicts an example of the CFA model for a hypothetical measure. The CFA model for the two groups resulted from the analyses conducted for

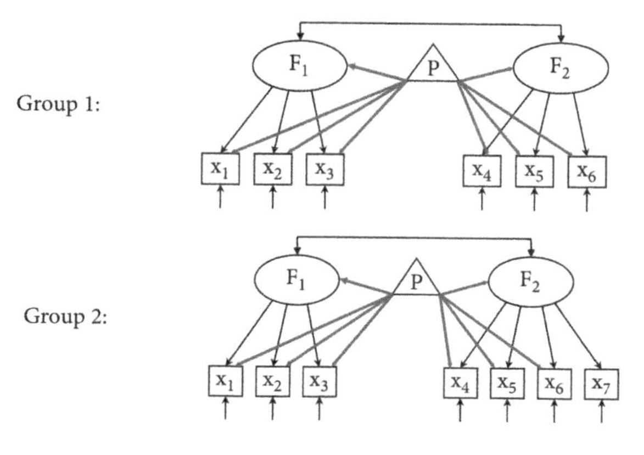

Figure 3.2 Model 1: Configural Equivalence

Model 0. In this example, results from Model 0 indicated the presence of partial-configural (pattern) equivalence across Groups 1 and 2 (see Figure 3.2). The pattern of fixed and free parameters does not have to be 100% identical for the equivalence testing to proceed. It is unclear where the cutoff should be, because no statistical guidelines have been established in the literature (Byrne, 2012). Theory and the potential impact of partial-configural equivalence on other forms of measurement validity (e.g., predictive validity) need to be considered when determining how much consistency between the two groups in the patterns of free and fixed parameters is needed. For both groups, the best-fitting model had two latent variables (F1 and F2). For Group 1, each factor was measured by three observed variables: F1 with X_1 to X_3, and F2 with X_4 to X_6. For Group 2, F2 had an additional indicator, X_7. The MG-CFA model was estimated with no constraints imposed among the factor loadings common to both models (see Figure 3.2). The loading associated with X_7 on F2 for Group 2 is not a part of the factor structure that is common to both groups; therefore, the loading cannot be used in the equivalence-testing process. The loading for X_7 is part of the MG-CFA analysis but not part of the equivalence-testing process.

Model 2: Weak Metric Equivalence

This model examines whether the construct validity of the measure is equivalent across groups. To test for construct-validity equivalency, the researcher has to examine unit equivalence or metric equivalence (Bryne, 2012; Meredith & Teresi, 2006). Weak metric equivalence is a test of the equality of the factor loadings (see Figure 3.3). Factor loadings are the regression slope linking the items to the latent variable. These regression slopes represent the expected change in the observed score on the change in the latent variable per unit for each item (Vandenberg & Lance, 2000). Equal factor loading across groups indicates that members of the two groups are interpreting the items in the same manner (Byrne, 1998). When weak metric equivalence is not present, some items might be more salient to the construct for one group than the other (Chen et al., 2010). Many consider the assessment of weak metric equivalence central to determining if a measure is "biased" (Millsap, 1989).

When weak metric equivalence has been established, the measurement units are identical across groups, but the origins of the scales can

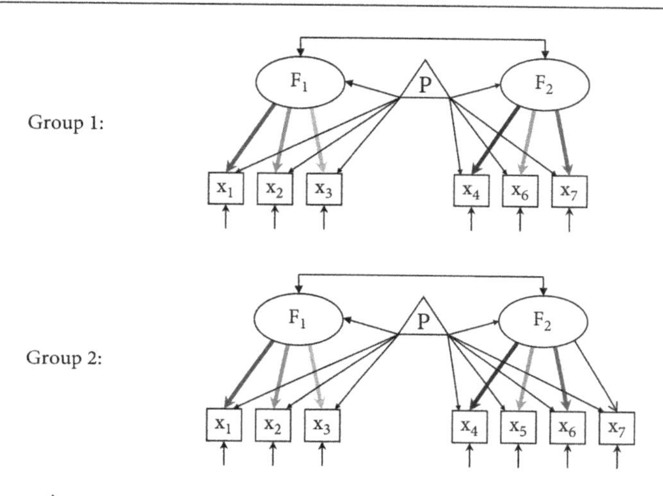

= Common Parameter colors indicate equivalency constraints imposed on factor loadings only. Residual variances & intercepts freely estimated.

Figure 3.3 Model 2: Weak Metric Equivalence

differ. For example, van de Vijver and Leung (1997) illustrated metric equivalency by using the Kelvin and Celsius scales. They observed that there is unit equivalency across the two scales, but also that the scales had origins that differed by approximately 273 degrees. Because the researcher knows the amount of difference between the two origins, he or she can make unbiased comparisons between the two scales. For example, the researcher knows that 0°C is equivalent to 273°K, and that both represent the point at which water freezes. Unfortunately, when we only have evidence of metric equivalency, we cannot assume equal origins for the scales between the two groups and can only make unbiased comparisons if we know the level of difference between the scales' origins (Vandenberg & Lance, 2000). Although evidence that no metric equivalence exists between groups supports the between-group construct validity of the measure, additional evidence of equivalency can be found by examining the measurement equivalence of other important parameters in the measure's factor structure, such as the factor variance and covariance, and residual covariances when present. Having established metric equivalence, which builds on the presence of configural equivalence, the researcher can then have confidence that the construct validity of the measure is equivalent across groups.

Model 3: Strong (Scalar) Metric Equivalence

This model investigates whether systematic response bias is the same across groups. Unfortunately, establishing metric equivalence does not allow researchers to make unbiased comparisons between the two groups (Meredith & Teresi, 2006; Yuan & Bentler, 2006). Scalar equivalence is needed for there to be unbiased comparisons between groups. Scalar equivalence involves imposing constraints on the factor loadings that are found to be equal during the metric analysis and on all factor intercepts common to both groups (see Figure 3.4). When scalar equivalence is found both groups have the same expected item response at the same absolute level of the trait being assessed. When scalar equivalence is found the observed differences in the items and scale means between the groups can be contributed to differences in the factor means. When there is evidence of scalar equivalence, the researcher can be confident that the manifest variables are measuring the same latent variables across groups. Additionally, when scalar equivalence is present, it is possible to make between-group comparisons based on the latent means. Without scalar equivalence, the differences in the latent group means will be confounded by differences in the manifest variable intercepts (Meredith & Teresi, 2006).

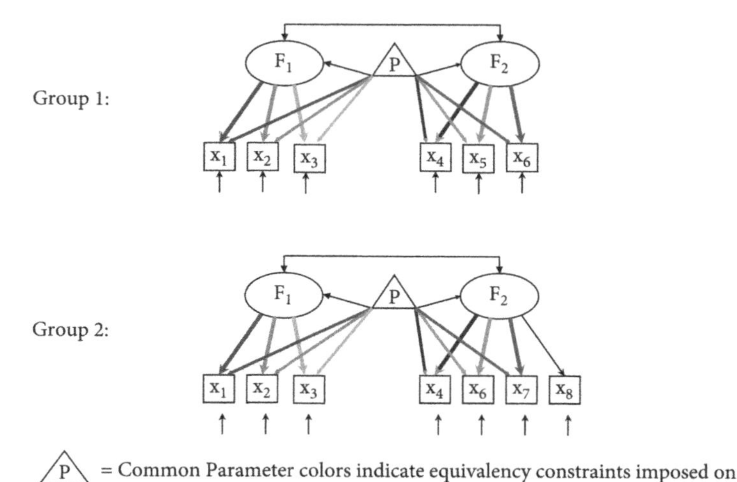

Figure 3.4 Model 3: Strong Metric (Scalar) Equivalency

Model 4: Strict Metric (Error Variance and Covariance) Equivalence

This model examines whether a comparable level of item reliability exists across the groups. Strict metric equivalence involves examining the equivalency of the residual variance, and, when present, residual covariances between the two groups. Figure 3.5 depicts the situation where only the residual variances are examined between the two groups. It has been argued that the presence of scalar equivalence does not guarantee that differences in the scale intercorrelations, observed means, and variances between groups are due solely to differences in the latent means and variance (DeShon, 2004; Meredith, 1993; Meredith & Teresi, 2006). Strong (scalar) metric equivalence does not ensure that error variances are equal (i.e., item and scale reliability). For instance, using both single-item and multi-item measures, DeShon demonstrated that differences in the heterogeneity of error variances (i.e., differences in reliability) will contribute to group differences. Whereas scalar metric equivalence requires constraints on all factor intercepts, strict metric equivalence requires that constraints also be placed on the variances of the residuals. Meredith and Teresi (2006) noted that, when strict equivalence holds true, the factor

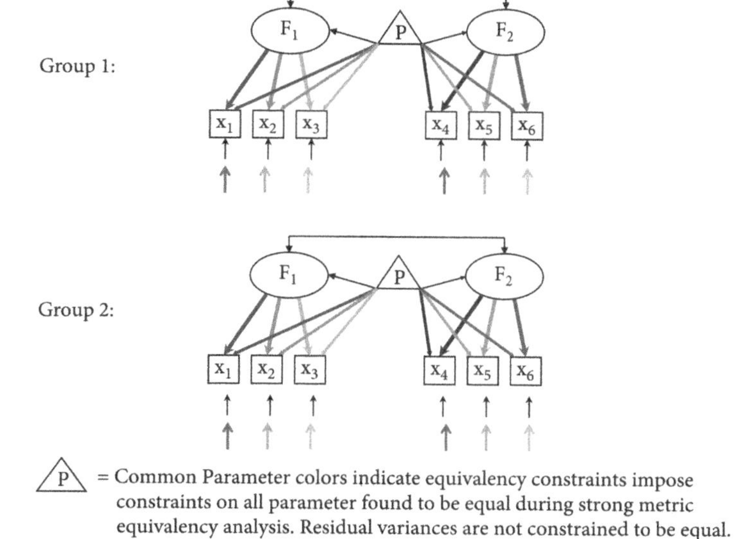

Figure 3.5 Model 4: Strict Metric Equivalency

mean and variance accounts for group differences in the manifest variables' means and variances. DeShon (2004) explained that measurement error, which is captured in the variance of the residuals for the manifest variables, is not solely made up of random processes, but incorporates unmodeled sources of systematic error that can have an impact on an individual's responses to measures. Establishing strict metric equivalence—or establishing the equivalence of the reliability of items that make up a measure—addresses the impact of those sources of systematic error on the observed variables.

In their re-analysis of the data used in the Tenijenhuis, Tolboom, Resing, and Bleichrodt (2004) study, Wicherts and Dolan (2010) provided support for the importance of assessing strict metric equivalence. In assessing the strong metric equivalence of a Dutch children's intelligence test, Tenijenhuis et al. concluded that this measure was appropriate for use with both Turkish and and Moroccan children. When conducting a strict metric equivalence analysis, Wicherts and Dolan demonstrated the inadequancy of strong metric equivalence. They found that at least three of the 12 subscales on the intelligence test were seriously biased toward seven-year-old children of Moroccan and Turkish decent. Researchers have called into question the utility of strict measurement invariance because many situations can affect it, and most research questions only require testing either weak or strong measurement equivalence (Byrne, 2012; Widaman & Reise, 1997). Scholars have argued that there are logical reasons why residual variance would not be equal across groups; for instance, some variance is generally expected between a sample and the larger population (Widaman & Reise, 1997).

Model 5: Equivalence of Factor Variance

This model examines whether there is equal heterogeneity in the latent variables between groups. Examining the measurement equivalence should focus not only on the comparability of the factor structure underlying the measure, but also on distributional aspects of the measure. In addition to the measurement equivalence of the mean structure, the factor (latent) variance should be examined. Weak metric equivalence tells researchers the extent to which an item's content may vary in salience across groups (Spini, 2003).

Model 6: Equivalence of Factor Covariance

Evaluating the equivalence of factor variance provides evidence about the equivalence of a structural parameter of the measure. Specifically, this test addresses the extent of homogeneity in the variance of common factor means between the study groups. In most cases, testing for the invariance of factor variances is not an important concern. The evaluation of the invariance of factor covariances and factor means is of greater concern. Invariance of factor variance is required to establish invariance in the item reliabilities (Vandenberg & Lance, 2000).

This model investigates the equivalence in the associations between factors across groups. Similar to examining the measurement equivalence of factor variance, testing for common factor covariance provides another opportunity to examine the equivalence of a structural aspect of the measure. The motivation for this test is to evaluate the consistency of the internal structure of the measure between groups.

Model 7: Equivalence of Latent Means

This model investigates whether there are differences between the latent means of the study groups. Two different types of means can be derived when administering a measure: observed and latent. Observed means are calculated directly from the manifest variables or the raw data (Byrne, 2012). Latent means, as with latent variables, are not directly observable. Latent means are associated with the latent variables found in the measurement or structural model and are estimated from the observable manifest variables in the model (Bentler & Yuan, 1999). The desire to determine the measurement equivalence of a measure is often driven by a need to compare group means.

In many instances, strict metric equivalence cannot be obtained and, in general, is believed to be a restrictive requirement (Byrne, 2012); therefore, it is typical for researchers to establish measurement equivalence only at the level of strong metric equivalence. Once strong metric equivalence has been determined, researchers typically proceed with *t*-test or Analysis of Variance (ANOVA) to evaluate group differences (Sharma, Durvasula, & Ployhart, 2011). This method has been described as a piecemeal approach grounded in an ANOVA framework that fails to consider one important weakness: the observed means do not incorporate potential differences in the reliability of the measurement instrument used to obtain the observed mean. In fact, when using *t*-test or ANOVAs, we assume that the researcher is using measures that have a high degree of reliability and that there is no difference in the reliability of the measure between groups. When this is not true, any estimates of differences in means will be biased (Bobko, Roth, & Bobko, 2001; Sharma et al., 2011). Incorporating the mean structure into the measurement-equivalence analysis allows researchers to determine the measurement equivalence of the factor means or the latent group means (i.e., the measurement equivalence of the intercepts across groups).

Testing the measurement equivalence of the latent means builds directly on establishing strong (scalar) equivalence and bases the analysis on the MACS rather than merely on the COV of the measure. This strategy represents an integration of an approach that emphasizes the change of behavior between groups with an approach that looks at the association of variables within a group (Polyhart & Oswald, 2004). Testing the equivalence of the latent means allows the latent means for factors to be tested simultaneously while holding equivalent aspects of the measure and the structural part of the model equal (Polyhart & Oswald, 2004). Polyhart and Oswald (2004) pointed out that examining the measurement equivalence of latent means allows researchers to integrate a concern about group differences with a focus on the covariance of variables across individuals within the groups. A MACS-based measurement-equivalence analysis can also be used to examine the differential item functioning of individual items (for a more detailed discussion, please see Wu, 2009).

The measurement-equivalence of Models 1 to 3 is typically the primary concern of researchers seeking to establish the appropriateness of a measure across diverse groups because of the focus on parameters that make up the measurement aspect of the instrument. Models 4 and 5 address structural parameters that are important for establishing the strength of the measure when used within a comparative framework that focuses on group means. Models 6 and 7 focus on structural aspects of the measure that are less frequently addressed; in many instances, they are not relevant enough to the study at hand.

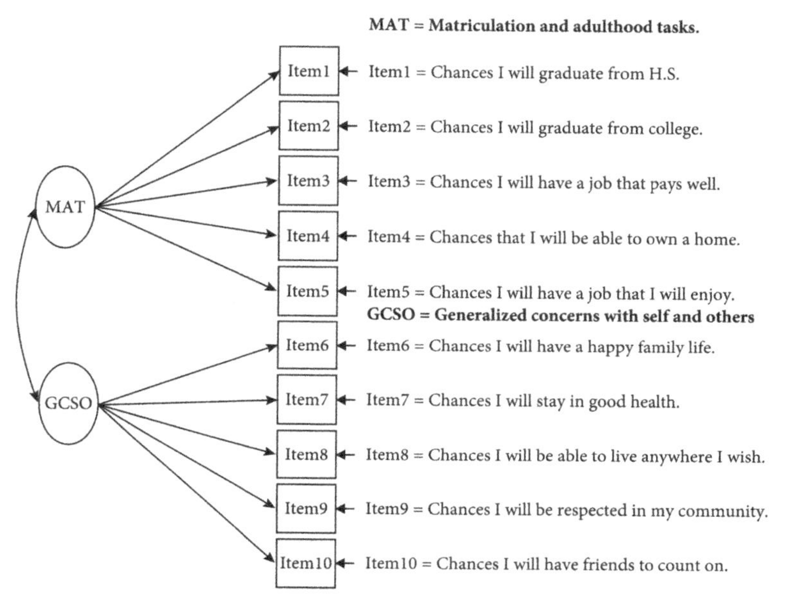

Figure 3.6 Measurement Model for the Measure of Participants' Future-Orientation Interests

ILLUSTRATION

This illustration examines the measurement equivalence of a measure of adolescents' future-orientation interests (Farmer, 2002). The measure was used in a student survey (McLaughlin & Lee, 1997) of the National Education Longitudinal Study of 1988 (NELS:88). The future-orientation-interest scale consists of two factors: The first factor Matriculation and Adulthood Tasks (MAT) contains five items related to matriculation and adulthood tasks, and the second factor Generalized Concerns with Self and Others (GCSO) contains five items that focus on general concern for oneself and others (see Figure 3.6. Farmer, 2002). In Farmer's (2002) study students responded to each item using a 5-point rating scale that ranged from 1 (*very low likelihood*) to 5 (*very high likelihood*). For the purpose of this illustration, only the African American and Hispanic tenth-graders in college preparation programs were included in this analysis. Additionally, the measurement equivalence between male ($N = 284$) and female ($N = 333$) students was investigated.

Distributional Analysis

The process of evaluating the equivalence of a measure begins with examining the distributional properties of the measure for each of the study groups. The goal is to determine if there is any evidence of potential bias, such as, for instance, acquiescence bias. Attention is given to the distribution of participants' responses across the response categories for each of the measure's items. Table 3.2 displays the response distribution across the gender groups from Farmer's (2002) study. Table 3.3 provides information on the items' skewness and kurtosis. Overall, for all the items, the distribution of responses appears to be similar between the two groups (see Table 3.2). As might be expected for adolescents attending a college-preparation high school program, they were generally optimistic about their future. Both the male and female participants' response categories were negatively skewed. For both groups greater negative skewness was found on items related to expected educational attainment (see Table 3.3). Male participants exhibited slightly more negative skewness than female participants (see Table 3.3). There appeared to be little evidence of differences in the distributions of responses across the two groups.

Baseline Measurement Models

Separate Group Analysis

Figure 3.6 provides the hypothesized measurment that was evaluated for each separately. The researcher's goal at this step in the analysis is to identify the appropriate measurement model for both groups separately. Parsimony, substantive meaningfulness, and statistical parameters were used to determine the best-fitting model. Due to the non-normal nature of the distribution of the items, a robust maximum likelihood with robust standard errors (MLR) estimate was used to implement non-normal robust standard error calculations (Muthén & Muthén, 1998–2007).

Hispanic and African American Males, Baseline Model Analysis

Results from the initial CFI analysis of the scale for the male participants resulted in an acceptable fit: MLR $\chi^2_{[34]} = 82.31, p < .05$; Scaling Correction Factors = 1.414; Log-likelihood$_{[31]} = -2559.03$; Scaling Correction Factor $_{MLR} = 1.40$; CFI = .95; RMSEA = .07 [CI 95%: .050, .089]; SRMR = .05. An examination of Sorbom's (1989) MI and the Expected-Parameter-Change (EPC) statistic (Saris, Satorra, & Sorbom, 1987) is used here to indicate

Table 3.2 Response Distribution Across Gender

	Frequencies									
	Males (*n* = 284)					*Females* (*n* = 333)				
Item	*1*	*2*	*3*	*4*	*5*	*1*	*2*	*3*	*4*	*5*
Chances that you will graduate H.S.	0.0	0	3.5	15.8	80.7	0.2	.2	2.4	13.4	83.9
Chances that you will be able to go to college.	0.9	1.4	11.9	29.1	56.6	0.6	1.2	9.6	25.7	62.9
Chances of you having a job that pays well.	0.0	.2	16.1	36.0	47.7	0.0	.2	.6	13.2	51.9
Chances that you will be able to own home.	0.5	2.3	17.0	31.7	48.5	0.6	1.8	18.5	30.6	48.5
Chances you will have a job you enjoy.	0.2	1.6	16.7	31.0	50.5	0.4	1.0	11.2	36.0	51.4
Chances you will have a happy family life.	0.9	1.2	17.2	34.8	45.9	1.0	1.2	14.6	39.4	43.8
Chances you will stay in good health.	0.2	1.4	16.7	35.4	46.2	0.2	2.2	23.3	39.6	34.7
Chances you will be able to live anywhere you wish.	2.4	4.7	25.2	31.4	36.3	0.8	4.7	24.5	34.7	35.3
Chances you will be respected in the community.	0.2	1.2	21.0	40.9	36.6	0.0	1.2	17.8	45.7	35.2
Chances you will have friends to count on.	0.7	1.2	16.0	38.2	43.9	0.6	1.8	16.2	40.9	40.5

Note: Items were rated on a 5-point rating scale where 1 = *very low likelihood* and 5 = *very high likelihood*.

Table 3.3 Skewness and Kurtosis Across Gender

	Male (n = 284)		*Female (n = 333)*	
	Skewness	*Kurtosis*	*Skewness*	*Kurtosis*
1. Chances that you will graduate H.S.	−2.124	3.739	−3.036	11.841
2. Chances that you will be able to go to college.	−1.395	1.959	−1.573	2.499
3. Chances of you having a job that pays well.	−.708	−.268	−.918	.223
4. Chances that you will be able to own home.	−.921	.239	−.895	.203
5. Chances you will have a job you enjoy	−.883	−.008	−1.090	1.117
6. Chances you will have a happy family life.	−.968	.790	−1.35	1.289
7. Chances you will stay in good health.	−.779	−.067	−.439	−.544
8. Chances you will be able to live anywhere.	−.726	.035	−.578	−.307
9. Chances you will be respected in the community	−.484	−.422	−.424	−.563
10. Chances you will have friends to count on.	−.906	.726	−.830	.558
Mean of skewness and kurtosis values	−0.989	0.6723	1.113	1.632

places where the re-specification of the model should be considered. The MI value indicates the amount of change that would occur in the model's chi-square value if a given constrained parameter were to be estimated without constraints. To be conservative in this post hoc model re-specification, the minimum MI value was set to 10, which corresponds to $p < .01$ for a chi-square distribution with one degree of freedom. Additionally, the feasibility of the parameters' estimates and the appropriateness of the standard errors were used to determine whether a change in the model should be made. Determining the feasibility of the parameters involved examining whether the parameters fell outside the admissible ranges (e.g., correlations greater than 1.00 or negative variances). To avoid merely capitalizing on chance, model revisions were based on both the statistical and substantive meaningfulness of the revision.

The MI and EPC statistics indicate that three changes should be made to the model: (1) Item 5 (chances you will have a job that you enjoy) should load on the second factor (general concern for self and others), (2) Item 6 (chances you will have a happy family life) should load on the first factor (matriculation and adulthood tasks), and (3) a correlation was indicated between the residual variances for Item 1 (chances I will graduate from high school) and Item 2 (chances I will graduate from college). Making the modifications indicated above to the model resulted in two new models. The first model included the cross-loading of Item 5 on the second factor. The MI indexes indicated that including the cross-loading between Items 1 and 2 would improve the model fit. The second model included both the cross-loading of Item 5 on Factor 2 and the correlation between the residual variances of Items 1 and 2. The cross-loading of Item 6 (chances you will have a happy family life) on Factor 1 (matriculation and adulthood tasks) did not remain statistically significant once these two items were added to the model.

In the final CFA model (see Figure 3.7), Item 5 (chances you will have a job that you like) was found to cross-load on the second factor (generalized concern with self and others). Additionally, covariance was observed between Item 1 (chances I will graduate from high school) and Item 2 (chances I will graduate from college). Table 3.4 contains the model-fit information for both the initial CFA measurement-model and the revised (final) model. The CFA's model-fit information for the revised baseline model was MLR $\chi^2_{[33]} = 56$, $p < .05$; Scaling Correction Factors = 1.37; Log-likelihood$_{[41]}$ = −2597.82; Scaling Correction Factor MLR = 1.40; CFI = .98; RMSEA = .05 [CI 95%: .03, .07]; SRMR = .03. In addition to the inspection of the improvement in fit statistics between the initial and revised CFA measurement models, an adjusted chi-square difference test was carried out (see Table 3.4). The adjusted chi-squared difference test is used to determine if there is a statistically significant difference in the model fit between the constrained and unconstrained models. When the imposed model constraints are warranted there will

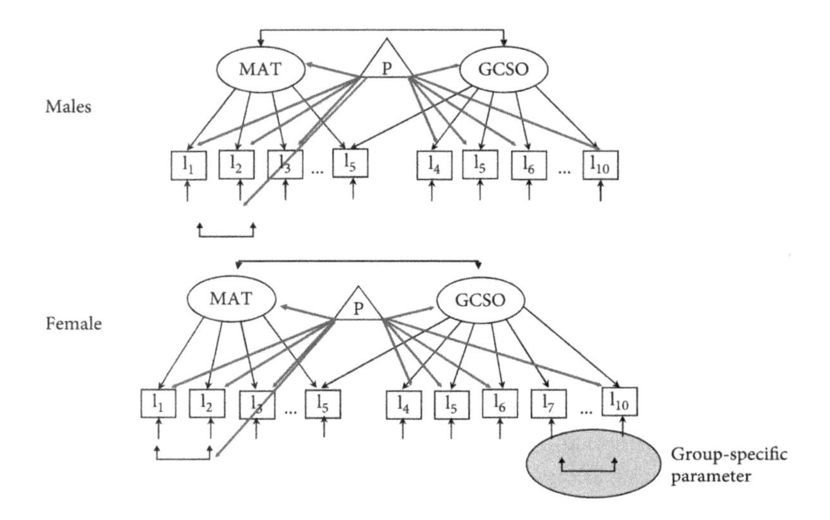

Note: MG-CFA analysis, MAT = Matriculation and adulthood, and GCSO = Generalized concerns with self and others.
P –parameters common to both groups which will be freely estimated between the groups.

Figure 3.7 Base CFA Measurement Model for Males and Females

Table 3.4 Baseline Model Fit Information

Model	χ^2_{MLR}	*df*	*SC*[a]	*CFI*	*RMSEA*	*Difference Test*	
						$\Delta\chi^2_{Adjusted\ MLR}$	*df*
Male							
Initial	82.31	34	1.41	.95	.07 (.05, .09)		
Final	56.44	32	1.37	.98	.05 (.03, .07)	18.90**	2
Female							
Initial	91.98	34	1.28	.94	.07 (.05, .09)		
Final	42.68	31	1.22	.99	.03 (CI: .00, .06)	34.56**	3

[a]SC = Scale Correction factor for maximum likelihood robust parameter estimates with standard errors and a chi-square test statistic (MLR); CFI = Comparative Fit Index; RMSEA = Root Mean Square Error of Approximation, df = degree of freedom.
** $p < .01$.

not be a significant difference found when comparing the chi-squares for nested models. Satorra and Bentler's (2010) adjusted chi-square test is needed in this case because the typical chi-square difference test is not appropriate when robust parameter estimations are used. When using an estimator appropriate for data that has a non-normal distribution like MLR, an adjusted chi-square difference test is needed to compare differences in chi-squares for nested models (Satorra, 2000). The procedures for carrying out the adjusted chi-square difference test are described on the Mplus website (StatModel.com). Appendix A presents the procedures used to calculate the adjusted chi-square difference test for the initial and revised CFA measurement models. The results of the adjusted chi-square difference test indicate a significant difference between the initial and revised models (ΔMLR $\chi^2_{(2)} = 18.90, p < .01$; see Table 3.4).

Hispanic and African American Females, Baseline Model Analysis
The results of the initial CFI analysis (see Figure 3.6) indicated that Item 5 (chances you will have a job that you like) cross-loaded on the second factor (general concern for self and others). Additionally, there was covariance between Item 1 (chances I will graduate from high school) and Item 2 (chances I will graduate from college). The model-fit information for the revised baseline model was MLR $\chi^2_{(33)} = 56, p < .05$; Scaling Correction Factors = 1.37; Log-likelihood$_{(41)}$ = −2597.82; Scaling Correction Factor MLR = 1.40; CFI =.98; RMSEA = .05 [CI 95%: .030, .070]; SRMR = .03.

The results from the initial CFI analysis (see Figure 3.6). of the measure for the female participants resulted in an acceptable fit: MLR $\chi^2_{(51)} = 91.98, p < .05$, Scaling Correction Factor MLR = 1.41; CFI = .94; RMSEA = .07 [CI 95%: .054, .089]; SRMR = .05. Consistent with the findings for the male participants, the EPC for the female participants indicated the need to cross-load Item 5 on the second factor (general concern for self and others) and correlate the residual variances for Items 1 and 2. The EPC also indicated that the correlation between the residual variance for Item 6 (chances I will have a happy family life) and Item 7 (chances I will stay in good health) would significantly improve the model's fit. The CFA's model-fit information for the revised baseline model (see Figure 3.7) was MLR $\chi^2_{(31)} = 42.68, p < .05$, Scaling Correction Factors MLR = 1.22; Scaling; CFI = .99; RMSEA = .03 [CI 95%: .000, .060]; SRMR = .03. The results of the adjusted chi-square difference test are presented in Table 3.4 and indicate significant difference between the initial and revised models (adjusted $\Delta\chi^2_{(2)} = 34.56, p < .01$).

Figure 3.7 depicts the baseline model for the Hispanic and African American male and female participants. Except for the addition of a correlation between the residual variance for Items 6 and 7, the configurations of the patterns of free and fixed parameters were identical. Because of the difference in the number of correlated residuals variances between the two groups, only partial measurement equivalence can be tested. The next step in the process is to carry out the MG-CFA, which will directly test the configural measurement equivalence of the scale.

Multi-Group Confirmatory Factor Analysis

Model 1: Test of Configural Measurement Equivalence

In this MG-CFA analysis, all model parameters were freely estimated simultaneously between the groups. Figure 3.8 shows the parameters that are freely estimated between the two groups. The correlation between the residual variances for Items 6 and 7 is only present for Hispanic and

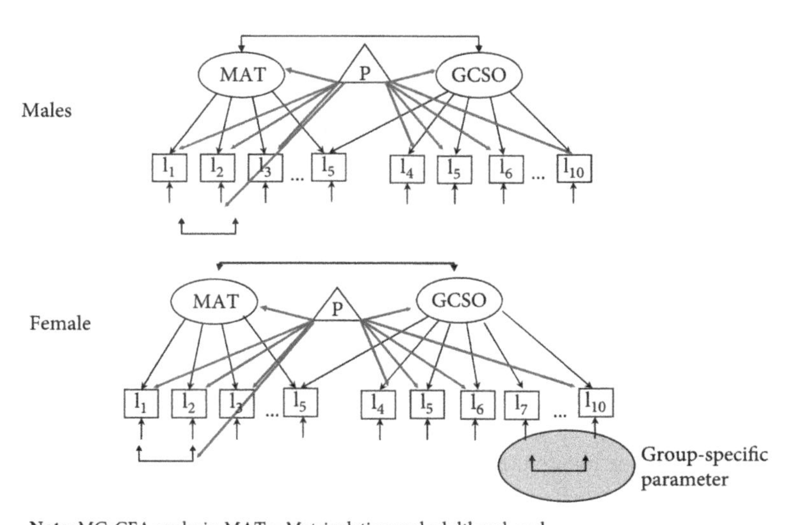

Note: MG-CFA analysis, MAT = Matriculation and adulthood, and GCSO = Generalized concerns with self and others.
P –parameters common to both groups which will be freely estimated between the groups.

Figure 3.8 Model 1: Configural Equivalence; Multi-Group Confirmatory Factor Analysis; Male and Female, Future Orientation Interests

African American females and therefore was only estimated for these groups; consequently, only a test of partial measurement equivalence can be carried out. Because the two groups were tested simultaneously, this MG-CFA allowed for a direct test of the multi-group measurement equivalence of the measure. The fit information for this model was also compared to the multi-group model, which involved constraining the factor loadings to determine measurement equivalence of these parameters (i.e., the weak metric equivalence model). Table 3.5 contains the abbreviated Mplus 7 program syntax and the results for the configural model. Although a detailed discussion of the Mplus 7 programming language is beyond the scope of this text, we will explain the aspects of the program that are related to examining the configural model. Readers interested in a more detailed discussion of the Mplus 7 programming language should consult the Mplus 7 user guide (Muthén & Muthén, 1998–2007), which is available online (at http://statmodel.com/ugexcerpts.shtml). Additionally, Mplus 7 (http://statmodel.com/) has available an extensive library of free online training video on the use of the program. Detailed examples of how to implement an MG-CFA to evaluate measurement equivalence are available for other structural equation programs, such as IBM® SPSS® Amos (Byrne, 2004), EQS 6.2 (Byrne, 2008), and LISREL 9.1 (Fleishman & Benson, 1987).

Lines 1 to 5 identify the location, format, and structure of the data file. Please be aware that the line numbers are not a part of the programming language, they have been placed there for easy reference to the commands being discussed. The data file contained the 10 items (Items 1 to 10: see Table 3.5, Line 2) that made up the measure. The missing values in the data set are identified in Line 3. Because both groups are contained in the same data file, a command is needed to identify the two groups (see Table 3.5, Line 4). *Gender* is the grouping variable used to identify the two groups (coded 1 for males and 2 for females). In this illustration, with the exception of a covariance, the configuration of the measurement model for the reference group (males) is the same as that found in the comparison group (females). The command that identifies the location, format, and structure of the data file was the same for all analyses and therefore will not be repeated when the sequential models are discussed.

As was the case with the separate group models, the MLR estimate with standard errors and chi-square statistics that are robust to violations of normality was used (see Table 3.5, Line 6). There are two sets of model commands (see Table 3.5, Lines 7 to 13 and Lines 14 to 23) needed to identify the multi-group measurement models that were estimated. By default the MODEL command imposes equality constraints across the two groups on two of the measurement parameters, factor loadings and intercepts of continuous indicators (or thresholds for categorical indicators). Structural parameters, factor variances and covariances, and factor means will not be constrained to be equal across groups by default. The ways to constrain and free parameters across groups will be illustrated.

The first set of model commands (see Table 3.5, Lines 7 to 12) identifies the parameters of the latent-variable structure common to both groups, which we refer to as the *common measurement model.* All of the measurement-equivalence models will contain the model's basic setup commands (see Table 3.5, Lines 7 to 12). These model parameters were the focus of the measurement equivalence testing. Based on the partial-configural equivalence results, MAT was made up of Items 1 to 5 (see Table 3.5, Line 7). The BY command assigned Items 1 to 5 to the first factor. The label assigned to the parameter appears in the parentheses next to the variable (i.e., indicator) name. For example, the factor loading for Item 1 (chances that you will graduate from high school) is named LOAD1.

The second factor GCSO is made up of six items: Items 5–10 (see Table 3.5, Line 8). To identify the model and to ensure that the factors are on the same scale for both groups, the same indicator for each factor was fixed when specifying the model in the common model command (MODEL) and in the model-specific commands (MODEL FEMALE). The WITH command is used to indicate the covariance between variables. The covariance between Items 1 and 2 is the only covariance common to both groups (see Table 3.5, Line 11). Because in this instance the analysis will be based on the covariance matrix (COV) and does not include the mean structure, which will be addressed later, the factor means were fixed to zero (Line 12). Brackets around the factor name [MAT] are used to indicate the factor mean. The at sign "(@)" is used to fix a parameter to a specific value. Command Line 12 [MAT-GCSO@0)] fixes the factor mean for MAT and GCSO to zero. As will be demonstrated later, the measurement equivalence of the mean structure can be incorporated into the analysis of the measure. By default, Mplus 7 freely estimates the intercepts and residual variances between the groups; therefore, no specification regarding these parameters is needed. The commands in Table 3.5 Lines 9 and 10 were included to make explicit that the intercepts and residual variances are freely estimated when carrying out a configural equivalence analysis.

The next set of model commands (see Table 3.5, Lines 13-17) are used to specify the group-specific measurement model and free the parameters that are needed to test the configural equivalence of the measure. These are the commands that will be changed when the other measurement equivalence models are tested. Because the baseline model for the male participants contained all of the same parameters present in the model common to both groups, no group-specific (Line 13) parameter specification is needed for this group. Lines 14, 15, and 16 under the female model specification free the parameters (factor loadings) common to both groups. Mplus 7 constrains these factor loadings to be equal between the two groups by default. For identification purposes, the first item for each factor (Item 1) was fixed to 1. Line 17 provides the specification of the factor variance and means for the group specific model. The factor variance was freely estimated (MAT-GCSO*) and the factor means were fixed to zero (MAT-GCSO@0).

The item name in brackets, [ITEM1] identifies the item intercept. Placing an asterisk (*) next to the name frees the intercept. The command on Line 18 frees the intercepts, which otherwise would have been constrained to be equal by the program's defaults. The command on Line 19 frees the residual variances between the groups.

The residual covariance common to both groups that was specified in Line 19 was freely estimated between the groups. The one residual covariant specific to the female sample had to be included in this group-specific model command (Line 20). Line 23 requests that Mplus 7 provide model descriptive information for the residuals (RESIDUAL), standardized and unstandardized factor loadings ("STAND(ALL)"), sample statistics (SAMPSTAT) and modification indexes ("MOD(3.84)"). The default critical value for modification indexes (MI) is 10. The MI critical value was reduced to 3.84, which corresponds to a p value of. 05. The results from the MI analysis will aid in evaluating the fit of the configural model. The MI analysis will help to identify whether there are parameters that are missing but should be included in the multi-group model.

The model-fit information for the configural model was MLR $\chi^2_{[62]} =$ 106.660, $p < .01$, Scaling Correction Factors = 1.308, Scaling Correction Factor MLR = 1.308; CFI = .978; RMSEA = .048 [CI (95%) = 032, .063]. Unlike when a maximum likelihood (ML) estimation method is used for an MG-CFA analysis, an MLR estimation does not always result in a chi-square for the multi-group model that is the sum of

Table 3.5 Model 1: Configural Equivalence Model Testing In Mplus 7

```
!COMMON MEASUREMENT MODEL SPECIFICATION!
  1. DATA:
     FILE IS 'G:\NELS_1988_92\CH.dat';

  2. VARIABLE:
     NAMES ARE ITEM1 ITEM2 ITEM3 ITEM4 ITEM5 ITEM6
     ITEM7 ITEM8 ITEM9 ITEM10 GENDER;

  3. MISSING ARE ALL (-99);

!COMMAND THAT IDENTIFIES THE TWO GROUPS!

  4. GROUPING IS GENDER (1 =MALE 2 = FEMALE);

!IDENTIFIES VARIABLES THAT WILL BE USED IN THE MODEL!

  5. USEVARIABLE are ITEM1 ITEM2 ITEM3 ITEM4 ITEM5 ITEM6
     ITEM7 ITEM8 ITEM9 ITEM10;

!ESTIMATOR ROBUST ML TO ADDRESS NON-NORMAL OF THE INDICATORS!

  6. ANALYSIS:

     ESTIMATOR = MLR;

  7. MODEL: MAT BY ITEM1@1 (LOAD1)1 ITEM2* (LOAD2) ITEM3*
    (LOAD3) ITEM4* (LOAD4); ITEM5* (LOAD5);

  8. GCSO BY ITEM6@1 (LOAD6) ITEM7* (LOAD7)ITEM8* (LOAD8)
     ITEM9* (LOAD9)
     ITEM10* (LOAD10) ITEM5* (LOAD5);

!LOADI INTERCEPTS—ALL FREE!

  9. [ITEM1*] (I1); [ITEM2*] (I2); [ITEM3*] (I3); [ITEM4*]
     (I I4); [ITEM5*] INTERCEPTI5); [ITEM6*] (I6);
     [ITEM8*] (I8); [ITEM9*]
     (I9); [ITEM7*] (I7);
     [ITEM10*] (I10);
!RESIDUAL VARIANCE—ALL FREE!

  10. ITEM1* (E1); ITEM2* (E2); ITEM3* (E3); ITEM4* (E4);
      ITEM5* (E5); ITEM6* (E6);

     ITEM8* (E8); ITEM9* (E9); ITEM7* (E7); ITEM10* (E10);
!COVARIANCE BETWEEN ERROR TERMS!

  11. ITEM1 WITH ITEM2*;
!FACTOR VARIANCE IS FIXED TO 1 FOR IDENTIFICATION PUROSES!
!FACTOR MEAN IS 0 (REQUIRED BY MPLUS 7.0)!
```

Table 3.5 *Continued*

```
  12. MAT-GCSO*; [MAT@0 GCSO@0];
!GROUP SPECIFIC MODEL!

  13. MODEL MALE:

  14. MODEL FEMALE:
!FACTOR LOADINGS—FREE BETWEEN THE GROUP!

  15. MAT BY ITEM1@1 ITEM2*
ITEM3* ITEM4* ITEM5*;

  16. GCSO BY ITEM6@1 ITEM7* ITEM8* ITEM9*
      ITEM10* ITEM5*;
!FACTOR VARIANCE FREE - FACTOR MEAN =0!

  17. MAT-GCSO*; [MAT@0];, [GCSO@0];
! ITEM INTERCEPTS—ALL FREE!

  18. [ITEM1*]; [ITEM2*]; [ITEM3*];
      [ITEM4*]; [ITEM5*]; [ITEM6*];
      [ITEM8*]; [ITEM9*]; [ITEM7*]; [ITEM10*];
! RESIDUAL VARIANCE—ALL FREE!

  19. ITEM1-ITEM10*;
!GROUP SPECIFIC COVARIANCE!

  20. ITEM1 WITH ITEM2*;
!ABSENCE OF MODEL SPECIFICATION CONSTRAINED LOADINGS EQUAL!
!BASED IN PREVIOUS ANALYSIS TWO INTERCEPTS FREED ALL OTHER
  CONSTRAINTS THE EQUAL!

  21. ITEM7 WITH ITEM8*;
!FACTOR VARIANCE FREE - FACTOR MEAN =0!

  22. MAT-GCSO*; [MAT-GCSO@0];
!OUTPUT PRINTS OUT REGRESSION WEIGHT STANDARDIZATION LOADS
  LOADINGS AND SAMPLSTAT!

  23. OUTPUT: RESIDUAL STAND(ALL) SAMPSTAT MOD(3.84);
```

Note: Illustration data from the National Educational Longitudinal Study of 1988. Examined differences between males (N = 284) and females (N = 333) on the self-report measure of youth's future orientation interests. This analysis was based on Farmer (2000) study.
Words within parentheses are the parameter label. For example, "ITEM1* (LOADI1)" labels the loading for ITEM1 as LOADI1.
The numbers next to the lines of Mplus syntax are numbered for reference purposes only. They are not a part of the Mplus programming language.

the chi-square found the in the single-group CFA analysis (Byrne, 2011). This model fit information serves as the foundation for testing the measurement equivalence of the measurement model. Having established support for the comparability of the configuration of the measurement model between the two groups, the next step is to examine the measurement equivalence of the factor loadings.

Model 2: Test of Weak Metric Measurement Equivalence

Figure 3.9 depicts the MG-CFA model, and Table 3.6 contains the Mplus 7 syntax that was used to examine the potential metric equivalence between the two groups. The syntax to provide the data file's specification information will be the same as it was for the configural measurement equivalence analysis and therefore will not be explained again here. Only the equivalence of the factor loadings will be tested. As can be seen in Figure 3.9, the measurement models for the two groups reflect only partial-configural equivalence. For the African American and Hispanic females, group-specific residual covariance exists between Item 7 and Item 8; therefore, only partial measurement equivalence can be tested. In Table 3.6, Lines 1 to 7 provide the Mplus 7 syntax specification for the

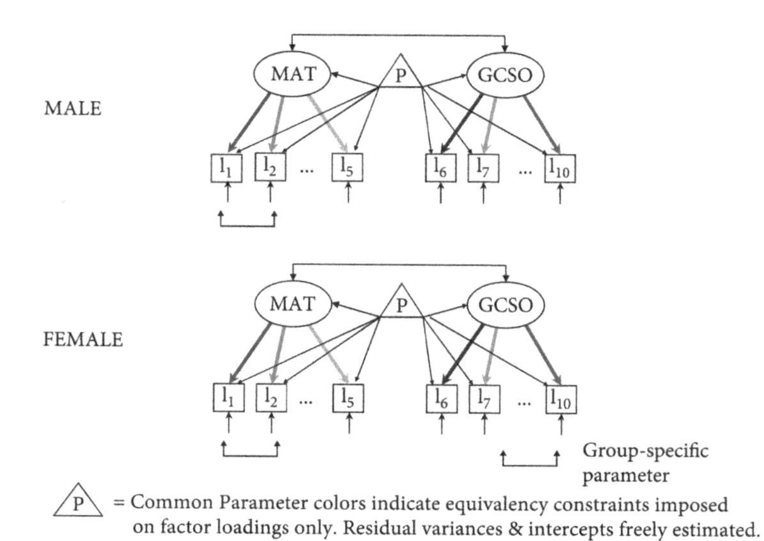

Figure 3.9 Model 2: Metric Equivalence Model; Multi-Group Confirmatory Factor Analysis; Male and Female, Future Orientation Interests

Table 3.6 Model 2: Weak Metric Equivalence Model Testing in Mplus 7

Factor loadings are constrained to be equal between the two groups.

```
!COMMON MEASUREMENT MODEL SPECIFICATION !
 1. ANALYSIS: ESTIMATOR = MLR;

 2. MODEL: MAT BY ITEM1* (LOADI1)
    ITEM2* (LOADI2);ITEM3* (LOADI3)
    ITEM4* (LOADI4); ITEM5* (LOADI5);

 3. GCSO BY ITEM6* (LOADI6) ITEM7* (LOADI7)
    ITEM8* (LOADI8) ITEM9* (LOADI9)
    ITEM10* (LOADI10) ITEM5* (LOADI5);
!LOADING INTERCEPTS - ALL FREE!

 4. [ITEM1*] (I1); [ITEM2*] (I2); [ITEM3*] (I3);
    [ITEM4*] (I I4); [ITEM5*] INTERCEPTI5);
    [ITEM6*] (I6);[ITEM8*] (I8); [ITEM9*] (I9);
    [ITEM7*] (I7);
    [ITEM10*] (I10);
!RESIDUAL VARIANCE - ALL FREE !

 5. ITEM1* (E1); ITEM2* (E2);
    ITEM3* (E3); ITEM4* (E4);
    ITEM5* (E5); ITEM6* (E6);
    ITEM8* (E8); ITEM9* (E9);
    ITEM7* (E7); ITEM10* (E10);
!COVARIANCE BETWEEN ERROR TERMS!

 6. ITEM1 WITH ITEM2*;
!FACTOR VARIANCE IS FREE TO 1 FOR IDENTIFICATION PUROSES!
!LOAD MEANS CONSTRAINED TO ZERO - MEAN STRUCTURE NOT
  A PART OF THE MODEL!

 7. MAT-GCSO@1; [MAT- GCSO@0];
!GROUP SPECIFIC MODEL !
!NO SPECIFICATION NEEDED - NOT GROUP SPECIFIC !

 8. MODEL MALE:

 9. MODEL FEMALE:
! LOAD LOADINGS - ALL EQUALITY BETWEEN THE THREE GROUP!

10. MAT BY ITEM1* (LOADI1) ITEM2* (LOADI2)
    ITEM3* (LOADI3) ITEM4* (LOADI4)
    ITEM5* (LOADI5);
```

(continued)

Table 3.6 *Continued*

Factor loadings are constrained to be equal between the two groups.

```
11. GCSO BY ITEM6* (LOADI6) ITEM7* (LOADI7)
    ITEM8* (LOADI8) ITEM9* (LOADI9)
    ITEM10* (LOADI10) ITEM5* (LOADI5);

!FREE LOAD MEAN!
!LOAD VARIANCE -AND MEAN FIXED TO 0!

12. MAT-GCSO*; [MAT-GCSO@0];

!ITEM INTERCEPTS - ALL FREE !

13. [ITEM1-ITEM10*];

!RESIDUAL VARIANCE - ALL FREE !

14. ITEM1- ITEM10*;

!GROUP SPECIFIC COVARIANCE !

15. ITEM1 WITH ITEM2*;

16. ITEM7 WITH ITEM8*;

!OUTPUT PRINTS OUT REGRESSION WEIGHT STANDARDIZATION LOADS
  LOADINGS AND SAMPLSTAT!

17. OUTPUT: RESIDUAL STDYX SAMPSTAT MOD(3.84);
```

Illustration data from the National Educational Longitudinal Study of 1988. Examined differences between males (N = 284) and females (N = 333) on the self-report measure of youth's future orientation interests. This analysis was based on Farmer (2000) study.
Words within parentheses are the parameter label. For example, "ITEM1* (LOADI1)" labels the loading for ITEM1 as LOADI1.
The numbers next to the lines of Mplus syntax are numbered for reference purposes only. They are not a part of the Mplus programming language.

measurement model that is common to both groups. Parameter labels were specified for this model; for example, the factor loadings for Items 1 to 3 were labeled LOAD1 to LOAD3. Because no model-specific parameters were used for males, no model specification was needed for males (see Line 8). In Lines 9 and 10, the factor loading between the two groups was constrained to be equal. By including the labels specified in the common model (see Lines 2 and 3) in the specification of the female-specific model (see Lines 9 and 10) the loadings were constrained between the two groups. The evaluation of the extent to which constraining the factor loadings to be equal between the two groups improves the fit of the multi-group model when compared to the configural model, and the results of the MI test will be used to determine the ME of the factor loadings. The fit statistics indicate a well-fitting model, CFI = .98, RMSEA = .04 [CI 95%: .020, .060], adjusted $\chi^2_{(73)} = 115.721$, MLR Scale Correction factor = 1.316. The weak metric equivalence model did not result in a significant decrease in the fit relative to the configural equivalence model (adjusted $\Delta\chi^2_{(11)} = 3.541$, $p > .05$).

Mplus 7.0 identifies all model parameters that would have to be altered to improve the model's fit. Table 3.7 contains the results of the modification indices. Only the modification indices related to the equivalency constraints for the factor loadings common to the two groups were

Table 3.7 Weak Metric Invariance Model Modification Indices

Group	*Parameter*	*M.I.*	*E.P.C.*	*Standard E.P.C.*	*Std YX E.P.C.*
Males					
	Factor Loading				
	MAT BY ITEM6	8.55	0.16	0.16	0.20
	Covariance				
	ITEM5 WITH ITEM1	4.30	−0.03	−0.03	−0.15
	ITEM5 WITH ITEM4	7.03	0.04	0.04	0.26
	ITEM8 WITH ITEM5	4.85	0.06	0.06	0.17
	ITEM9 WITH ITEM8	4.41	0.06	0.06	0.18
	MATS64J WITH ITEM8	5.03	−0.08	−0.08	−0.17
	MATS64J WITH ITEM9	6.17	0.06	0.06	0.20
Females					
	Covariance				
	ITEM4 WITH ITEM3	9.02	−0.07	−0.07	−0.47
	ITEM7WITH ITEM6	9.44	0.09	0.09	0.24
	ITEM8 WITH ITEM6	7.26	−0.09	−0.09	−0.22
	ITEM9 WITH ITEM3	4.11	0.03	0.03	0.17
	ITEM9 WITH ITEM8	5.58	0.07	0.07	0.21
	ITEM10WITH ITEM9	4.94	0.06	0.06	0.17
Females					
	Factor Loadings				
	GCSO BY ITEM2	4.012	−0.226	−0.122	
	Covariances				
	ITEM4 with ITEM3	5.302	−0.053	−0.341	
	ITEM7 with ITEM6	12.291	0.100	0.272	
	ITEM8 with ITEM6	5.128	−0.070	−0.183	
	ITEM9 with ITEM3	4.922	0.035	0.186	
	ITEM9 with ITEM7	6.186	−0.068	−0.241	

M.I. = Modification index; E.P.C. = expected parameter change index; Std E.P.C. = Standardized expected parameter change index examined. These findings indicate that constraining the factor loadings to be equal across the groups was supported. Overall, the results provide evidence of the equivalency of the factor loadings across the African American and Hispanic male and female youth.

Model 3: Test of Strong (Scalar) Metric Measurement Equivalence

Building on the findings from the weak metric measurement equivalence analysis, the equivalency of the intercepts for the indicators with equivalent factor loadings was investigated. Figure 3.10, depicts the MG-CFA measurement model that was estimated. Table 3.8 contains the Mplus 7 syntax corresponding to that model. The syntax for the model-specific measurement model for females (Lines 7 to 15) constrained the factor loadings (see Table 3.8, Lines 9 and 10) and intercepts (see Table 3.8, Line 12) between the two groups. The labels on the factor loadings and intercepts found on the common and group specific measurement model constrained those parameters' loadings between the groups. The overall fit for the model was good (CFI = .97, RMSEA = .048, [CI: 95%: .035, .062]); but the results from the modification indices indicated that two item intercepts were not equal between the two groups. The MI information for Item 7 (chances you will stay in good health) for males was MI = 12.69, EPC = .09, $p < .05$, whereas the female MI information for the item was MI 12.69; EPC = −12.69, $p < .05$. The equivalency constraint for Item 7's intercept was relaxed, and the model was re-estimated. The fit information for the model indicated good model fit: CFI = .98, RMSEA = .04, [CI (95%): .025, .055]; $\chi^2_{[80]}$ = 125.156, Scaling Correction Factor = 1.286. The post hoc MI analysis indicated that equivalency constraint for Item 10's (chance I will have friends to count on) intercept should be relaxed. The MI information for the males for Item 10 was MI = 4.022, EPC = .056, $p < .05$; for Female Item 7, MI = 4.021; EPC = −.072, $p < .05$). The model was estimated with the equivalency constraints for the intercepts for Items 7 and 10 relaxed. The fit information for the model indicated good model fit: CFI =.98, RMSEA =.04, [CI 95%: =.020,.050]; $\chi^2_{[79]}$ = 120.726, Scaling Correction Factor = 1.290). The MI information did not indicate that any other parameters constrained to be equal across the groups needed to be relaxed. The partial strong metric equivalence measurement model did not result in a significant decrease in the fit relative to the weak metric equivalence model (adjusted $\Delta\chi^2_{[6]} = 3.541$, $p > .05$). In the model with the equality constraint for all the items, the intercept was 4.192 for Item 7 and 4.221 for Item 10.

Table 3.8 Model 3: Strong Metric (Scalar) Invariance in Mplus 7

Intercepts constrained to be equal between the two groups.

```
!COMMON MEASUREMENT MODEL SPECIFICATION !
 1. MODEL: MAT BY ITEM1* (LOADI1) ITEM2* (LOADI2);
    ITEM3* (LOADI3) ITEM4* (LOADI4); ITEM5* (LOADI5);

 2. GCSO BY ITEM6* (LOADI6) ITEM7* (LOADI7)
    ITEM8* (LOADI8) ITEM9* (LOADI9)
    ITEM10* (LOADI10) ITEM5* (LOADI5);
!LOADI INTERCEPTS - ALL FREE !

 3. [ITEM1*] (I1); [ITEM2*] (I2); [ITEM3*] (I3); [ITEM4*]
    (I I4);
    [ITEM5*] INTERCEPTI5); [ITEM6*] (I6);
    [ITEM8*] (I8); [ITEM9*] (I9); [ITEM7*] (I7);
    [ITEM10*] (I10);
!RESIDUAL VARIANCE - ALL FREE !

 4. ITEM1* (E1); ITEM2* (E2); ITEM3* (E3); ITEM4* (E4);
    ITEM5* (E5); ITEM6* (E6);
    ITEM8* (E8); ITEM9* (E9);
    ITEM7* (E7); ITEM10* (E10);
!COVARIANCE BETWEEN ERROR TERMS!

 5. ITEM1 WITH ITEM2* (ED_COV);
!FACTOR VARIANCE IS FREE TO 1 FOR IDENTIFICATION PUROSES!
!LOAD MEANS CONSTRAINED TO ZERO - MEAN STRUCTURE NOT
  A PART OF THE MODEL!

 6. MAT-GCSO@1; [MAT@0 GCSO@0];
!GROUP SPECIFIC MODEL!
!NO SPECIFICATION NEEDED - NOT GROUP SPECIFIC !

 7. MODEL MALE:

 8. MODEL FEMALE:
!LOAD LOADINGS - ALL EQUALITY BETWEEN THE THREE GROUP!

 9. MAT BY ITEM1* (LOADI1) ITEM2* (LOADI2)
    ITEM3* (LOADI3) ITEM4* (LOADI4)
    ITEM5* (LOADI5);

10. GCSO BY ITEM6* (LOADI6) ITEM7* (LOADI7)
    ITEM8* (LOADI8) ITEM9* (LOADI9)
    ITEM10* (LOADI10) ITEM5* (LOADI5);
!FREE LOAD MEAN!
!FACTOR VARIANCE AND MEAN FREED!
```

(continued)

Table 3.8 *Continued*

Intercepts constrained to be equal between the two groups.

```
11. MAT-GCSO*; [MAT-GCSO*];
! ITEM INTERCEPTS - CONSTRAINED TO BE EQUAL !

12. [ITEM1*] (I1); [ITEM2*] (I2);
 [ITEM3*] (I3); [ITEM4*] (I I4);
 [ITEM5*] (I5); [ITEM6*] (I6); [ITEM7*] (I7);
 [ITEM8] (I8) [ITEM9](I9); [ITEM10*] (I10);
! RESIDUAL VARIANCE - ALL FREE !

13. ITEM1- ITEM10*;
! GROUP SPECIFIC COVARIANCE!

14. ITEM1 WITH ITEM2*;

15. ITEM7 WITH ITEM8*;
!OUTPUT PRINTS OUT REGRESSION WEIGHT STANDARDIZATION LOADS
  LOADINGS AND SAMPLSTAT!

16. OUTPUT: RESIDUAL STDYX SAMPSTAT MOD(3.84);
```

Note: Illustration data from the National Educational Longitudinal Study of 1988. Examined differences between males (N = 284) and females (N = 333) on the self-report measure of youth's future orientation interests. This analysis was based on Farmer (2000) study.
Words within parentheses are the parameter label. For example, "ITEM1* (LOADI1)" labels the loading for ITEM1 as LOADI1.
The numbers next to the lines of Mplus syntax are numbered for reference purposes only. They are not a part of the Mplus programming language.

For males, the item intercept (or the item factor mean) for Item 7 was 4.263 when the equality constraint on the item was eased, and 4.256 for Item 10. For females, the item intercept for Item 7 was 4.030 when the equality constraint on this item was eased, and 4.128 for Item 10.

Model 4: Test of Strict Metric (Error-Variance and Covariance) Equivalence

Error variance constraints Having established the partial equivalency of the intercepts (or partial strong metric equivalence) of the measure, it is now possible to examine the error-variance equivalence of the items across the groups. It should be kept in mind that, until the equivalence of the factor variance has been established, the equivalence of the error variance will not provide evidence of the equality of the item's reliability (Vandenberg & Lance, 2000). Because common error covariance is present, the equality constraint for this parameter will need to be tested. Table 3.9 provides Mplus 7 syntax for this model. The changes to the syntax needed to test the strict

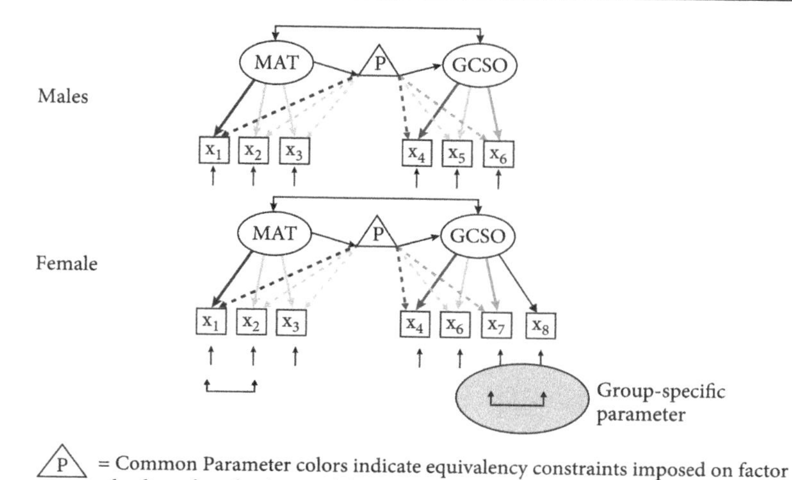

= Common Parameter colors indicate equivalency constraints imposed on factor loadings found to be equal during weak metric equivalency analysis & intercepts contrained to be equal. Residual variances are freely estimated.

Figure 3.10 MG-CFA Measurement Model

measurement equivalence begin on Line 9. Lines 9 and 10 constrain the factor loadings between the two groups. Line 12 constrains the intercepts, and Line 13 constrains the residual variances. Based on the results from the scalar metric analysis, no equality constraints for the intercept or the residual variance were imposed on Item 7 (chances you will stay in good health) or Item 10 (chances I will have friends to count on). The results of the analysis indicated that removing the equality constraints on Item 3 (chances I will have a job that pays well) and Item 6 (chances I will have a happy family life) residual variances would improve the model's fit. The fit information for the partial strict metric equivalence model that contained relaxed equality constraints for Items 7 and 10 item intercepts is CFI = .98, RMSEA = .04, [CI 95%: .024, .053]; $\chi^2_{[87]} = 129.826$, Scaling Correction Factor = 1.382. The fit information for the partial strict metric equivalence model that contained relaxed equality constraints for Items 3, 6, 7, and 10 residual variances is CFI = .98, RMSEA = .03, [CI 95%: .020, .051]; $\chi^2_{[85]} = 120.418$, Scaling Correction Factor = 1.351. There was a significant reduction in the adjusted change in chi-square ($\Delta\chi^2$) between the two models (adjusted $\Delta\chi^2_{[6]} = 6.242$, $p > .05$). For items where the residual variance equivalence held, the amount of item variance not accounted for by the factors was the same between the two groups. This was not the case for Item 3 (chances I will have a job that pays well), Item 6 (chances I will have a happy family life), Item 7 (chances

Table 3.9 Model 4: Strict Metric (Error-Variance and Covariance) Invariance in Mplus 7

Residual variance constrained to be equal between the two groups.

```
!COMMON MEASUREMENT MODEL SPECIFICATION !

 1. MODEL: MAT BY ITEM1* (LOADI1) ITEM2* (LOADI2);
    ITEM3* (LOADI3) ITEM4* (LOADI4); ITEM5* (LOADI5);

 2. GCSO BY ITEM6* (LOADI6) ITEM7* (LOADI7)
    ITEM8* (LOADI8) ITEM9* (LOADI9)
    ITEM10* (LOADI10) ITEM5* (LOADI5);

!LOADI INTERCEPTS - ALL FREE!

 3. [ITEM1*] (I1); [ITEM2*] (I2); [ITEM3*] (I3); [ITEM4*]
   (I I4);
    [ITEM5*] INTERCEPT(I5); [ITEM6*] (I6);[ITEM8*] (I8);
    [ITEM9*] (I9); [ITEM7*] (I7); [ITEM10*] (I10);

!RESIDUAL VARIANCE - ALL FREE !

 4. ITEM1* (E1); ITEM2* (E2);
    ITEM3* (E3); ITEM4* (E4); ITEM5* (E5); ITEM6* (E6);
    ITEM8* (E8); ITEM9* (E9); ITEM7* (E7); ITEM10* (E10);

!COVARIANCE BETWEEN ERROR TERMS!

 5. ITEM1 WITH ITEM2* (ED_COV);

!FACTOR VARIANCE IS FIXED TO 1 FOR IDENTIFICATION PURPOSES !
!LOAD MEANS CONSTRAINED TO ZERO - MEAN STRUCTURE NOT
A PART OF THE MODEL!

 6.  MAT-GCSO@1; [MAT@0]; [GCSO@0];

!GROUP SPECIFIC MODEL!
!NO SPECIFICATION NEEDED - NOT GROUP SPECIFIC !

 7. MODEL MALE:

 8. MODEL FEMALE:

!LOAD LOADINGS - ALL EQUALITY BETWEEN THE TWO GROUPS!

 9. MAT BY ITEM1* (LOADI1) ITEM2* (LOADI2)
    ITEM3* (LOADI3) ITEM4* (LOADI4)
    ITEM5* (LOADI5);

10. GCSO BY ITEM6* (LOADI6) ITEM7* (LOADI7)
    ITEM8* (LOADI8) ITEM9* (LOADI9)
    ITEM10* (LOADI10) ITEM5* (LOADI5);

!FREE LOAD MEAN!
!FACTOR VARIANCE AND MEAN FREED!
```

Table 3.9 *Continued*

Residual variance constrained to be equal between the two groups.

```
11. MAT-GCSO*; [MAT-GCSO*];
!ITEM INTERCEPTS - CONSTRAINED TO BE EQUAL - EXCEPT
  ITEM 3. ITEM 6, ITEM 7 AND ITEM 10 !

12. [ITEM1*] (I1); [ITEM2*] (I2); [ITEM3*]; [ITEM4*]
  (I I4); [ITEM5*] (I5); [ITEM6*]; [ITEM7*]; [ITEM8]
  (I8) [ITEM9](I9); [ITEM10*];

! RESIDUAL VARIANCE - CONSTRAINED TO BE EQUAL - EXCEPT
  ITEM 3. ITEM 6, ITEM 7 AND ITEM 10 !

13. ITEM1* (E1); ITEM2* (E2); ITEM3*; ITEM4* (E4);
    ITEM5* (E5); ITEM6*; ITEM8* (E8); ITEM9* (E9);
    ITEM7*; ITEM10*;

! CONSTRAINED FACTOR COVARIANCE TO EQUAL !.

14. ITEM1 with ITEM2* (ED_COV);

! GROUP SPECIFIC COVARIANCE!

15. ITEM1 WITH ITEM2*;

16. ITEM7 WITH ITEM8*;

!OUTPUT PRINTS OUT REGRESSION WEIGHT STANDARDIZATION
LOADINGS AND SAMPLE STATISTICS

17. OUTPUT: RESIDUAL STDYX SAMPSTAT MOD(3.84);
```

Note: Illustration data from the National Educational Longitudinal Study of 1988. Examined differences between males (N = 284) and females (N = 333) on the self-report measure of youth's future orientation interests. This analysis was based on Farmer (2000) study.
Words within parentheses are the parameter label. For example, "ITEM1* (LOADI1)" labels the loading for ITEM1 as LOADI1.
The numbers next to the lines of Mplus syntax are numbered for reference purposes only. They are not a part of the Mplus programming language.

you will stay in good health), or Item 10 (chances I will have friends to count on).

Common error covariance To test the equality of the covariance between the residuals, the group-specific measurement model syntax was altered to indicate that there would be an equality constraint for the residual covariance between the two groups. The original command that allowed the covariances to be estimated freely was on Line 14 (ITEM1 WITH ITEM2*) was altered and the new Line 14 is ITEM1 with ITEM2* (ED_COV). The addition of the label for the covariance statement to the model-specific set of commands for female will constrain this parameter to be equal between the two groups. The results from the modification indices did not indicate that the constraint on the residual covariance needed to be relaxed. The fit information for the model is as follows: CFI = .98, RMSEA = .036, [CI 95%: .020, .051]; $\chi^2_{[86]}$ = 121.692, Scaling Correction Factor = 1.361. There was not a significant reduction in the adjusted change in chi-square between the model with residual variance constraints and the residual covariance model (adjusted $\Delta\chi^2_{[1]} = 1.326$, $p < .05$).

Summary of the Results Assessing Measurement Equivalence

The following presentation of the results for measurement equivalence is written up as it would be for a publication.

The evaluation of the measurement equivalence of the Future Orientation Interest measure between African and Hispanic American males and females provided evidence of partial strict (metric) equivalence. Table 3.10 provides a summary of the areas of equivalence found when the parameters making up the measurement model were examined. For Items 7 (chances you will stay in good health) and 10 (chances you will have friend to count on), there was evidence of non-equivalence of the intercepts. For these items the level of systematic response bias for African and Hispanic American male and female tenth-graders in the sample cannot be assumed to be equal. For items 3, 6, 7, and 10, there was also a lack of evidence of equivalence of the residual variance.

Table 3.11 provides the model fit information for the four models that were examined. Overall, it can be argued that there was evidence of measurement equivalence at the strict (metric) level of equivalence. In most instances this level of measurement equivalence would be enough to utilize the measure with African and Hispanic American male and female tenth-graders. Having examined the measurement parameters that make up the measure, we will now examine the structural parameters.

Model 5: Equivalence of Factor Variance

The extent of the heterogeneity of the latent variable is tested when the equivalence of factor variances is examined. Lines 1 to 8 establish the configuration of the measurement model that is common to both groups (see Table 3.12). Mplus 7.0 requires that the factor means be fixed to 0 (See Line 3). Identification of the model will be achieved by fixing the factor variances to 1 in the overall model (See Lines 4 and 5). The constraints for the factor loadings (Lines 11 and 12), intercepts (Line 16), residual variances (Line 17) and residual covariances (Line 18) remain

Table 3.10 Areas of Measurement Invariance

Item	*Factor*		*Residual*	
	Loading	*Intercepts*	*Variance*	*Covariance*
1. Chances that you will graduate H.S.	X	X	X	X
2. Chances that you will be able to go to college.	X	X	X	
3. Chances of you having a job that pays well.	X	X	—	
4. Chances that you will be able to own home.	X	X	X	
5. Chances you will have a job you enjoy	X	X	X	
6. Chances you will have a happy family life.	X	X	—	
7. Chances you will stay in good health.	X	—	—	
8. Chances you will be able to live anywhere you wish.	X	X	X	
9. Chances you will be respected in the community	X	X	X	
10. Chances you will have friends to count on.	X	—	—	

Note: 'X' indicate a lack of invariance across the male and females study participants. "—" indicates invariance across males and females study participants.

from Model 4. The absence of labels for the intercepts for Items 7 and 10 allows these parameters to be freely estimated (Line 16). The absence of labels for the residual variances for Items 3, 6, 7 and 10 allows these parameters to be freely estimated between the two groups (Line 17).

The comparison of the change in the adjusted chi-squares between the partial strict metric equivalence model [adjusted $\chi^2_{[85]} = 120.418$, Scale Correction factor = 1.352] and the factor variance equivalence model [adjusted $\chi^2_{[88]} = 124.852$, Scale Correction factor = 1.348] was insignificant [adjusted $\Delta\chi^2_{[3]} = 4.452, p < .05$]. The fit information for the model was CFI = .98, RMSEA = .037, [CI 95%: .020, .051]. These results provide evidence that African American and Hispanic male and female tenth-graders had equal levels of heterogeneity of the latent variable, future-interest orientation.

Table 3.11 Model Fit Information Base CFA Models, Configural, Metric, Scalar and Residual Variance Models

Invariance Models	X^2_{MLR}	*df*	SC^+	*CFI*	*RMSEA (95% C.I.)*	*Difference Test* ΔX^2 *Adjusted MLR*	Δdf
(1) Configural	106.660	62	1.308	.978	.048 (.032, .063)		
(2) Weak Metric	115.721	73	1.316	.979	.043 (.028, .058)	9.361	11
(3) Partial Strong (Scalar) Metric	120.726	79	1.290	.979	.041 (.025, .055)	3.541	6
(4) Partial Strict Metric	120.418	85	1.352	.982	.036 (.020, .051)	3.260	6

Note: + SC=Scale Correction factor for maximum likelihood parameter estimates with standard errors and a chi-square test statistic (MLR); CFI = Comparative Fit Index; RMSEA = Root Mean Square Error of Approximation, Δdf = Difference is Degrees of Freedom

Model 6: Equivalence of Factor Covariance

Testing of the equivalence of the factor covariance builds on the findings from the previous invariance testing results. In Mplus 7.0 factor loadings and intercepts are constrained equal by default across groups and residual variances are estimated by default. In order to make it explicit, in the previous analysis, equality constraints were imposed by labeling parameters in both the general model command (for example, see Table 3.12, Line 1) and one of the group specific model commands (for example, see Table 3.12, Line 11). In this analysis, Mplus 7.0 program defaults were used to impose equality constraints for the factor loadings and intercepts. The absence of the specification of factor loadings in the group specific model indicates that the default to impose constraints for these parameters (See Lines 8 and 9) will be used. The presence of the intercepts for Items 7 and 10 in the group specific model for females (See Lines 11 and 12) will free the intercepts for Items 7 and 10 between the groups. The specification and labeling of the residual variances in both the general model (See Line 3) and in the group specific model for females (See Line 12) will constraint the residual variance for those items containing labels in both command lines. In this case the residual variances for Items 3, 6, 7 and 10 were not constrained equal between the groups.

The fit information for the model is as follows: CFI = .98, RMSEA = .036, [CI 95%: 019,050]; adjusted $\chi_{2\ [89]}$ = 125.071, Scale Correction factor = 1.345. The results from the modification indices did not indicate

Table 3.12 Model 5: Factor Variance Invariance Testing in Mplus 7

Factor variance constrained to be equal between the two groups.

```
!REFERENCE GROUP FACTOR VARIANCE INVARIANCE MODEL!
!FACTOR LOADINGS - ALL FREE!
 1. MODEL: MAT BY ITEM1* (LOAD1); ITEM2* (LOAD2); ITEM3*
    (LOAD3)*; ITEM4* (LOAD4)*; ITEM5* (LOAD5);
 2. GCSO BY ITEM6* (LOAD6); ITEM7* (LOAD7); ITEM8* (LOAD8);
    ITEM9* (LOAD9); ITEM10* (LOAD10); ITEM5*(LOAD5);
!FACTOR MEAN (FACTOR MEAN = 0 - REQUIRED BY MPLUS 7)!
 3. [MAT@0];[GCSO@0];
 ! FACTOR VARIANCE - FIXED TO 1 - MODEL IDENTIFICATION
   PURPOSES !
 4. MAT@1;
 5. GSCO@1;
!INTERCEPTS FREE!
 6. [ITEM1*] (I1); [ITEM2*] (I2); [ITEM3*] (I13); [ITEM4*]
    (I4); [ITEM5*] (I5); [ITEM6*] (I6); [ITEM7*] (I7);
    [ITEM8*] (I8); [ITEM9*] (I9); [ITEM10*] (I10);
!RESIDUAL VARIANCE - ALL FREE!
 7. ITEM1* (E1); ITEM2* (E2); ITEM3* (E3); ITEM4* (E4);
    ITEM5* (E5); ITEM6* (E6); ITEM7* (E7); ITEM8* (E8);
    ITEM9* (E9); ITEM10* (E10);
!COVARIANCE BETWEEN ERROR TERMS - FREE!
 8. ITEM1 WITH ITEM2* (ED_COV);
!GROUP SPECIFIC MODEL INFORMATION!
 9. MODEL MALE:
!FACTOR LOADING CONSTRAINED EQUAL BETWEEN THE GROUPS!
10. MODEL FEMALE:
11. MODEL: MAT BY ITEM1* (LOAD1); ITEM2* (LOAD2); ITEM3*
    (LOAD3); ITEM4* (LOAD4); ITEM5* (LOAD5);
12. GCSO BY MAT BY ITEM6* (LOAD6); ITEM7* (LOAD7); ITEM8*
    (LOAD8); ITEM9* (LOAD9); ITEM10* (LOAD10); ITEM5*
    (LOAD5);
!FACTOR VARIANCE FREE - FACTOR MEAN FIXED TO 1 - TESTING
FACTOR INVARIANCE!
13. MAT@1;
14. GCSO@1;
```

(continued)

Table 3.12 *Continued*

```
!FACTOR MEAN - FREE
15. [MAT-GCSO*];
!INTERCEPTS ALL FIXED EQUAL EXCEPT ITEM 7 AND ITEM 10!
16. [ITEM1*] (I1); [ITEM2*] (I2); [ITEM3*] (I13); [ITEM4*]
    (I4); [ITEM5*] (I5); [ITEM6*] (I6); [ITEM7*]; [ITEM8*]
    (I8); [ITEM9*] (I9); [ITEM10*];
!RESIDUAL VARIANCES ALL EQUAL EXCEPT ITEM 7 AND ITEM 10!
17. ITEM1* (E1); ITEM2* (E2); ITEM3* (E3); ITEM4* (E4);
    ITEM5* (E5); ITEM6* (E6); ITEM7*; ITEM8* (E8);
    ITEM9* (E9); ITEM10*;
!COVARIANCE BETWEEN ERROR TERMS - CONSTRAINED EQUAL BETWEEN
THE TWO GROUPS!
18. ITEM1 WITH ITEM2* (ED_COV);
!GROUPS SPECIFIC COVARIANCE!
19. ITEM1 WITH ITEM2*;
20. ITEM7 WITH ITEM8*;
!OUTPUT PRINTS OUT REGRESSION WEIGHTS STANDARDIZED LOADINGS
AND SAMPLE STATISTICS.
21. OUTPUT: RESIDUAL STDYX SAMPSTAT MOD (3.84);
```

Note: Illustration data from the National Educational Longitudinal Study of 1988. Examineddifferences between males (N = 284) and females (N = 333) on the self-report measure of youth's future orientation interests. This analysis was based on Farmer (2000) study.
Words within parentheses are the parameter label. For example, "ITEM1* (LOADI1)" labels the loading for ITEM1 as LOADI1.
The numbers next to the lines of Mplus syntax are numbered for reference purposes only. They are not a part of the Mplus programming language.

Table 3.13 Model 6: Factor Covariance Invariance Testing in Mplus 7

Factor covariance constrained equal between the two groups

```
!REFERENCE GROUP FACTOR VARIANCE INVARIANCE MODEL!
!FACTOR LOADINGS - METRIC SET - FIRST ITEMS FIXED TO 1!
 1. MODEL: MAT BY ITEM1 (LOAD1); ITEM2* (LOAD2); ITEM3*
    (LOAD3)*; ITEM4* (LOAD4)*; ITEM5* (LOAD5);
 2. GCSO BY ITEM6 (LOAD6); ITEM7* (LOAD7); ITEM8* (LOAD8);
    ITEM9* (LOAD9); ITEM10* (LOAD10); ITEM5* (LOAD5);
!RESIDUAL VARIANCES ALL EQUAL!
 3. ITEM1* (E1); ITEM2* (E2); ITEM3* (E3); ITEM4* (E4);
    ITEM5* (E5); ITEM6* (E6); ITEM7* (E7); ITEM8* (E8);
    ITEM9* (E9); ITEM10* (E10);
```

Table 3.13 *Continued*

```
 4. MAT@0];[GCSO@0];
!FACTOR VARIANCE - PARMETER LABEL - CONSTRAINTS EQUAL!
 5. MAT(1);
 6. GSCO(2);
!COVARIANCE BETWEEN ERROR TERMS - FREE!
 7. ITEM1 WITH ITEM2* (ED_COV);
!GROUP SPECIFIC MODEL INFORMATION!
 8. MODEL MALE:
!FACTOR LOADING CONSTRAINED EQUAL BETWEEN THE GROUPS!
 9. MODEL FEMALE:
!FACTOR MEAN - FREE
10. [MAT-GCSO*];
!INTERCEPTS ALL FIXED EQUAL EXCEPT ITEM 7 AND ITEM 10!
11. [ITEM7];
12. [ITEM10];
!RESIDUAL VARIANCES ALL EQUAL EXCEPT ITEM 3, 6, 7 AND
ITEM 10!
13. ITEM1* (E1); ITEM2* (E2); ITEM3*; ITEM4* (E4);
    ITEM5* (E5); ITEM6*; ITEM7* ; ITEM8* (E8); ITEM9* (E9);
    ITEM10*;
!COVARIANCE BETWEEN ERROR TERMS - CONSTRAINED EQUAL!
14. ITEM1 WITH ITEM2* (ED_COV);
! GROUPS SPECIFIC COVARIANCE !
15. ITEM1 WITH ITEM2*;
16. ITEM7 WITH ITEM8*;
!OUTPUT PRINTS OUT REGRESSION WEIGHTS STANDARDIZED LOADINGS
AND SAMPLE STATISTICS.
17. OUTPUT: RESIDUAL STDYX SAMPSTAT MOD (3.84);
```

Note: Illustration data from the National Educational Longitudinal Study of 1988. Examined differences between males (N = 284) and females (N = 333) on the self-report measure of youth's future orientation interests. This analysis was based on Farmer (2000) study.
Words within parentheses are the parameter label. For example, "ITEM1* (LOADI1)" labels the loading for ITEM1 as LOADI1.
The numbers next to the lines of Mplus syntax are numbered for reference purposes only. They are not a part of the Mplus programming language.

that the constraint on the factor covariance needed to be relaxed. There was no significant reduction in the adjusted $\Delta\chi^2$ between the model with factor covariance constraints and the factor variance model (adjusted $\Delta\chi^2_{[1]} = .01$, $p < .05$). These results provided support for the equivalence of the factor covariances between the two groups.

Model 7: Testing for Latent Mean Invariance

Table 3.14 provides the Mplus syntax for testing for the latent mean invariance. Similar to testing for the equivalence of the factor variance and covariance, testing for the equivalence of latent means focuses on structural aspects of the instrument's measurement model. Before testing for the equivalence of the factor means, the researcher must have established that there is at least partial equivalence of the factor loading (weak metric equivalence) and intercepts (strong metric scalar equivalence). Due to model identification constraints, one of the groups is treated as a reference group, and differences between the factor means of the groups are tested (Byrne & Stewart, 2006). When estimating the MG-CFA model, the factor means for the reference group, in this case males, are fixed to zero (Table 3.14, see Line 7). Fixing the factor intercepts (i.e., factor means) to zero is also necessary because of the arbitrary nature of the origin, in those situations where the intercepts of the observed variables are constrained to be equal (Byrne, 2012). The estimated factor means in the comparison group represent the difference in the factor means between the reference and comparison groups (Byrne, 2012). Based on the results from the previous analyses, all of the factor loadings (see Table 3.15, Line 8, the model-specific command for females) were constrained to be equal across the groups. All of the intercepts, with the exception of the intercept for Items 7 and 10, were constrained to be equal across the groups (see Line 10). Also based on the results from analyses the residual variances for Items 3, 6, 7, 10 were not constrained equal between the two groups (See Line 11). To test for latent mean equivalence, the number of estimated intercepts must be lower than the number of measured variables. Equality constraints are used across the model to achieve this multi-group model identification requirement. In models where there is partial measurement equivalence and a large number of intercepts, the model may be under-identified. Our case example had two non-equivalent intercepts, which means that we had to estimate two intercepts for each group, making a total of four. Seven intercepts were found to be invariant;

Table 3.14 Model 7: Latent Mean Invariance Testing in Mplus 7

Factor means constrained to be equal between the two groups.

```
!COMMON MEASUREMENT MODEL SPECIFICATION!
!LOADING ALL FREE!
 1. MODEL: MAT BY ITEM1* (LOADI1) ITEM2* (LOADI2); ITEM3*
    (LOADI3)
    ITEM4* (LOADI4); ITEM5* (LOADI5);
 2. GCSO BY ITEM6* (LOADI6) ITEM7* (LOADI7); ITEM8* (LOADI8)
    ITEM9*
    (LOADI9); ITEM10* (LOADI10) ITEM5* (LOADI5);
!FACTOR VARIANCE AND MEAN !
 3. MAT@1; GCSO@1; [MAT-GCSO@0];
!INTERCEPTS - ALL FREE EXCEPT!
 4. [ITEM1*] (I1); [ITEM2*] (I2); [ITEM3*] (I3); [ITEM4*] (I4);
    [ITEM5*] (I5); [ITEM6*] (I6); [ITEM7*] (I7); [ITEM8*] (I8);
    [ITEM9*] (I9); [ITEM10*] (I10);
!RESIDUAL VARIANCE - ALL FREE!
 5. ITEM1* (E1); ITEM2* (E2); ITEM3* (E3);ITEM4* (E4);
    ITEM5* (E5); ITEM6* (E6);
    ITEM8* (E8); ITEM9* (E9); ITEM7* (E7); ITEM10* (i10);
!COVARIANCE BETWEEN ERROR TERMS!
 6. ITEM1 WITH ITEM2* (ED_COV);
!GROUP SPECIFIC MODEL INFORMATION!
MODEL MALE
!FACTOR MEAN AND VARIANCE - FIXED!
 7. F1-F2@1; [F1-F2@0];
 8. MODEL FEMALE:
    MODEL: MAT BY ITEM1* (LOADI1) ITEM2* (LOADI2);
    ITEM3* (LOADI3) ITEM4* (LOADI4); ITEM5* (LOADI5);
    GCSO BY ITEM6* (LOADI6) ITEM7* (LOADI7) ITEM8* (LOADI8)
    ITEM9* (LOADI9) ITEM10* (LOADI10) ITEM5* (LOADI5);
!FACTOR VARIANCE AND MEAN- FREE!
 9. MAT*; GCSO*; [MAT-GCSO];
!INTERCEPTS - CONSTRAINED EQUAL BETWEEN THE TWO GROUPS
EXCEPT ITEM 7 AND ITEM10!
10. [ITEM1*] (I1); [ITEM2*] (I2); [ITEM3*] (I3); [ITEM4*]
    (I4); [ITEM5*] (I5); [ITEM6*] (I6); [ITEM7]; [ITEM8*]
    (I8); [ITEM9*] (I9); [ITEM10*];
!RESIDUAL COVARIANCE - ALL FIXED EXCEPT ITEM3, ITEM6, ITEM7
AND ITEM10
```

(continued)

Table 3.14 *Continued*

```
11. ITEM1 (E1); ITEM2 (E2); ITEM3;
    ITEM4 (E4); ITEM5 (E5); ITEM6;
    ITEM7; ITEM8 (E8) ITEM9 (E9); ITEM10;

!COVARIANCE IS FIXED!

12. ITEM1 WITH ITEM2*; (ED_COV);

!GROUP SPECIFIC IS FREE!

13. ITEM7 WITH ITEM8;

14. OUTPUT: RESIDUAL STDYX SAMPSTAT MOD(3.84);
```

Note: Illustration data from the National Educational Longitudinal Study of 1988. Examined differences between males (N = 284) and females (N = 333) on the self-report measure of youth's future orientation interests. This analysis was based on Farmer (2000) study.
Words within parentheses are the parameter label. For example, "ITEM1* (LOADI1)" labels the loading for ITEM1 as LOADI1.
The numbers next to the lines of Mplus syntax are numbered for reference purposes only. They are not a part of the Mplus programming language.

therefore, 11 intercepts had to be estimated. None of the factor loadings were found to be invariant; therefore, 12 factor loadings had to be estimated. When the partial measurement equivalence model results in an under-identification of the multi-group model, it is recommended that researchers consider imposing additional equality constraints, which will result in fewer intercepts needing to be estimated (Bryne, 2012). In examining the results, the researcher should focus on the significance of the factor means for the comparison group, in this case the Hispanic and African American females. The factor means for MAT (Mean [*SE*] = .081[.09], $p > .05$) and GCSO (Mean [*SE*] = .031[.09], $p > .05$) were both non-significant. These results indicate that the factor means for the Hispanic and African American males and females were not significantly different.

The fit information for the model was CFI = .98, RMSEA = .036, [CI 95%: .019, .050]; adjusted $\chi^2_{[90]} = 126.082$, Scaling Correction factor = 1.340. The results from the modification indices did not indicate that the constraints on the factor covariance needed to be relaxed. There was not a significant reduction in the adjusted $\Delta\chi^2$ between the model with factor covariance constraints and the factor variance model (adjusted $\Delta\chi^2_{[2]} = .695$, $p < .05$). These results provided support for the equivalence of the factor mean between the two groups.

Summary of the Results Testing for Structural Equivalence

For presentation, the results of assessing structural equivalence are written up like they would be for a publication.

After finding evidence of partial strict metric measurement equivalence (or residual equivalence model), three structure parameters in the model were tested. First, the factor variance was constrained between the two groups, resulting in no significant decrease in the fit relative to the partial strict metric equivalence model (see Table 3.15, adjusted $\Delta\chi^2_{[3]} = 4.452$, $p < .05$). The results provide evidence that the African and Hispanic American male and female tenth-graders had very similar variability in their future orientation interests. Second, the factor covariances were constrained between the two groups and compared to the factor variance model. No significant decrease in the fit relative to the factor variance model was found (see Table 3.15, adjusted $\Delta\chi^2_{[1]} = .007$, $p < .05$). Finally, the latent mean variance model was estimated, and when compared to the factor covariance model, did not result in a significant decrease in the adjusted change in chi-square (see Table 3.15, adjusted $\Delta\chi^2_{[2]} = 0.695$, $p < .05$).

SUMMARY

Establishing measurement equivalence is an important process that must be done when conducting research with diverse groups. Assessing for measurement equivalence is a way of determining that the construct being measured has the same meaning across groups (configural equivalence); determining if the cutoff scores are the same across groups (scalar equivalence); and determining if the construct has the same dimensionality across groups as predicted by theory (Byrne & Vijver, 2010; structural equivalence).

In this chapter, we provided an overview of seven types of measurement equivalence, which are categorized into two types—measurement and structural. The types are configural, metric, scalar, error (i.e., covariance), factor variances, factor covariances, and factor means. As mentioned in Chapter 2, measurement equivalence can exist despite flaws in any phase of the research process. Therefore, we strongly urge the readers to use the processes and procedures detailed in Chapter 2 prior to assessing for measurement equivalence.

Multiple Group Confirmatory Factor Analysis was described as a statistical procedure that can be used to assess for measurement equivalence. The underlying assumptions of MG-CFA were reviewed. The steps necessary to evaluate the seven types of measurement equivalence were estimated, using MG-CFA in a case illustration. Mplus was used to analyze the data for the case illustration. The Mplus syntax for the MG-CFA analyses were provided. A write-up of the case example results was presented as if they are being reported in a publication.

In the next chapter, an additional hypothetical case example is presented to further illustrate the analytical procedures used to establish equivalence. This case example will also illustrate how descriptive statistics (means, standard deviations, skewness, and kurtosis) can be examined to determine initially if measurement equivalence has been established across the groups.

4

Hypothetical Case Illustration

OVERVIEW

In this chapter, we provide a hypothetical case to illustrate how to develop a research study while establishing equivalence at each phase of the research process.

HYPOTHETICAL CASE ILLUSTRATION

A researcher is interested in conducting a study that looks at racial and ethnic differences in depression among adolescents. In reviewing the literature, the researcher reads the article by Crockett et al. (2005) in which a CFA of the Center for Epidemiological Studies Depression Scale (CES-D) supported a 4-factor structure for non-Hispanic Caucasian adolescents and Mexican American adolescents, but not for Cuban American or Puerto Rican American adolescents. The four factors are Negative Affect, Positive Affect (reverse scored), Interpersonal Aspects, and Somatic Symptoms. The findings of the study suggested that the CES-D measures the same construct for non-Hispanic Caucasian adolescents and Mexican American adolescents, but not for Cuban American or Puerto Rican American adolescents.

Based on reading the Crockett et al. (2005) article, the researcher is able to make several decisions to ensure that the results of his or her own study reflect true group differences. Additionally, the literature helps the researcher choose an appropriate measure and decide what Hispanic group to include in the study. Because the results of the Crockett et al. study demonstrated that the CES-D measured the same constructs for both Mexican Americans and non-Hispanic Caucasians, the researcher decides to use the CES-D to assess depression and to include both groups of adolescents in the sample.

Selecting an appropriate measure that assesses the same construct for both Mexican American and non-Hispanic Caucasian adolescents is critical during the problem-formulation and measurement-selection phases of the research process. Recall that during both of these phases the researcher is concerned with establishing construct equivalence.

The researcher plans to use a non-experimental, comparative research design. A non-experimental design is needed because the independent variable (i.e., racial or ethnic group) cannot be manipulated by the researcher. A comparative research design is appropriate because the researcher is interested in looking at the differences in depression between two naturally occurring groups: Mexican American adolescents and non-Hispanic Caucasian adolescents.

To rule out alternative explanations for anticipated findings, the researcher has to address the threats to statistical-conclusion validity, internal validity, and construct validity. Regarding statistical-conclusion validity and construct validity, two aspects of construct equivalence (both conceptual/configural and metric) can be established by conducting an MG-CFA. Given that there may be group differences in response styles, which is a threat to internal validity, the researcher decides to assess whether the results are being affected by response styles by looking at the histograms, skewness, and kurtosis for each group separately. Because MG-CFA is being used for the data analysis, the researcher looks at the factor loadings and intercepts across groups to determine if there are group differences in response styles.

The researcher decides to use a convenience sample because the findings are not going to be generalized to the larger population of Mexican

American and non-Hispanic Caucasian adolescents. Although a convenience sample is used, the researcher can still enhance the comparability of the groups by collecting data on potential confounding variables (e.g., family socioeconomic status and gender), which can then be controlled for in the statistical analyses. Wanting to ensure that the results of the study can be attributed to true differences between the groups and not to the mode of data collection, the researcher decides to use only one mode of data collection: self-administered surveys.

In addition to establishing equivalence at each phase of the research process, equivalence still has to be established by statistical methods. The first step of the data analysis is to clean the data to make sure that the data were entered correctly. Any data that were entered incorrectly are deleted and replaced with the correct data.

The data preparation also involves an analysis of missing data. In our hypothetical example, let us assume that the results of the missing-value analysis indicate that cases with missing values are not systematically different from cases without missing values. A complete case approach is used, which is viewed as an acceptable approach when data are missing at random (Schaefer, & Graham, 2002). If the data are not missing at random, then imputing the missing values may be appropriate. (Researchers interested in learning more about imputing missing data, please refer to Baraldi and Enders [2010], Enders [2006], and Graham [2009]).

The second step is equivalence assessment. Equivalence assessment is done by examining the descriptive statistics for each group and establishing both conceptual/configural and metric equivalence of the CES-D, using an MG-CFA.

The researcher reviews the means, standard deviations, skewness, kurtosis, and histograms for each group separately. Histograms of the total score for each group and of the scores for each item of the CES-D for each group can be examined to determine the shape of the distributions for each group. Ideally, the histograms should show that these distributions of the total scores and the scores for the items of the CES-D are normally distributed. In the case example, the histograms showed that the total scores and the scores for the items were normally distributed. Based on an examination of these histograms, the researcher concludes that the distributions of the total scores of the CES-D for the Mexican American and non-Hispanic Caucasian adolescents are normal and similar in shape. If the histograms showed that the total scores and the scores for the items were not normally distributed and similar in shape, then the researcher would need to determine if the data were influenced by a particular response style. Differences in the distributions of the data may be initial indicators of there being differences in the reliability of a measure for the groups involved in the study prior to assessing the reliability of the measure (Tran, 2009). The researcher looks at the outliers for each of these groups separately to see if they fall within the same range for each group. Examining the outliers is also helpful in determining whether the respondents have a particular response style.

Conceptual/configural equivalence was assessed by conducting a CFA to determine whether the CES-D had the same factor structure for both study groups. Based on the work of Crockett et al. (2005), which suggested that a 4-factor model has an acceptable fit to the data for both groups, the researcher decides that this is the first model to test. As mentioned earlier, the four factors are Negative Affect, Positive Affect (reverse scored), Interpersonal Aspects, and Somatic Symptoms. The researcher decides to evaluate the model using the four indices discussed in chapter 3. The fit indices indicate that the 4-factor solution fits the data well for both groups—Mexican American adolescents and non-Hispanic Caucasian adolescents which provides evidence of conceptual/configural equivalence across the two groups.

Furthermore, the researcher can establish construct equivalence by assessing metric equivalence. Metric equivalence is established by imposing equality on the factor loadings across the two groups and fitting the factor model to the data for each group simultaneously. If the model fits well, metric equivalence is established. In the case example, the researcher establishes metric equivalence to determine whether the items on the CES-D are identical and have the same meaning across the two groups.

The metric equivalence of the measure is assessed by conducting an MG-CFA. The researcher compares a model in which the factor loadings are constrained to be equal across groups with one where the factor loadings are free to vary. A chi-square difference test is conducted to assess for significant differences between the constrained and unconstrained models. The factors are allowed to correlate freely in both models. In our example, the unconstrained 2-group model fits the data well. Additionally, the constrained model fits the data well. The chi-square difference test indicates that the unconstrained model is not significantly different from the constrained model, so the researcher has established full metric equivalence.

After determining that conceptual/configural and metric equivalence have been established, the researcher is interested in assessing the reliability of the measure. Given that this is a multidimensional measure, the researcher assesses its reliability using the appropriate procedures to calculate a multidimensional reliability coefficient. The multidimensional reliability coefficient is acceptable. Therefore, the researcher begins to analyze the data to test the study's hypothesis that Mexican American adolescents would be more depressed than non-Hispanic Caucasian adolescents, even after controlling for gender and family socioeconomic status. The researcher conducts a hierarchical multiple regression analysis. Gender and family socioeconomic status, which serve as the control variables, are entered at Step 1. Ethnicity is dichotomized (1 = *Mexican* and 0 = *non-Hispanic Caucasian*) and entered at Step 2. The final model is statistically significant. It is found that Mexican American adolescents are more likely to report being depressed than non-Hispanic Caucasian adolescents, even after controlling for gender and family socioeconomic status. The results are consistent with the researcher's hypothesis, and the researcher can be confident that the results are accurate because sufficient methodological precautions were taken to establish equivalence at each phase of the research process.

SUMMARY

We hope that the hypothetical case illustration provided the readers with a sense of how to establish equivalence at each phase of the research process. Additionally, the case illustrated how descriptive statistics (means, standard deviations, skewness, and kurtosis) can be examined to determine initially if measurement equivalence has been established across the groups.

In Chapter 5, which is the concluding chapter of this book, we will be discussing the following: the use of qualitative methods to establish measurement equivalence; the challenges of conducting research to establish equivalence using national datasets; and future directions for social work education at the doctoral level. Additionally, the contributions of this book to research methodology will be highlighted.

5

Conclusion

This chapter will focus on three topics: the use of qualitative methods to establish measurement equivalence; the challenges of conducting research to establish equivalence using national datasets; and future directions for social work education at the doctoral level.

QUALITATIVE METHODS IN ESTABLISHING MEASUREMENT EQUIVALENCE

The approaches to establishing measurement equivalence discussed in this book rely heavily on quantitative methods. Social work researchers conducting research with diverse groups need to be aware of the contribution that qualitative methods can make to establishing measurement equivalence. According to Knight et al. (2009), qualitative methods can be used to develop the measure to assess the construct of interest. Both focus groups and qualitative interviews can be used to develop items to be included on the survey. In other words, both focus groups and qualitative interviews can be used to establish conceptual equivalence, which is important at both problem-formulation and measurement-selection phases of the research process. In addition to noting that qualitative methods are helpful in establishing conceptual equivalence, Knight and colleagues also recommend that qualitative methods be used to detect non-equivalence. When quantitative methods fail to establish conceptual equivalence, focus groups or qualitative interviews can be conducted to determine if new items need to be added to the preexisting measures.

Given that researchers need to ensure that the construct has the same meaning across groups, employing both qualitative and quantitative methods will make an important contribution to our understanding of how persons conceptualize phenomenon under investigation. Therefore, we are recommending that content on the use of mixed methods to establish measurement equivalence be introduced in courses where social work doctoral students learn about conducting research with diverse groups.

THE CHALLENGES OF CONDUCTING RESEARCH TO ESTABLISH EQUIVALENCE USING NATIONAL DATASETS

National and large datasets are important resources that can be used by researchers to conduct research with diverse groups. Yet many of these datasets do not include the necessary information needed for researchers to establish equivalence when conducting research with these groups. This means that researchers using these datasets to look at differences between diverse groups may have inadvertently reported erroneous findings. Given this, it is critical that both national and large datasets include all the relevant information that will allow researchers to establish equivalence when conducting research with diverse groups. Information that needs to be included is the following: adequate sample sizes for all groups in the datasets so that researchers can conduct group comparisons that would allow the detection of statistically significant results; data at the individual level (i.e., data for each individual item on the scale and not merely the scale composite score); diverse groups identified by subpopulations so that researchers can account for within-group diversity; and contextual factors (e.g., geographical location, religious affiliation).

In datasets where data are not provided at the individual level, there needs to be documentation in the codebook that details how measurement equivalence was established so that researchers using these datasets can be assured that the results they obtained are reflective of true group differences. In addition to there being information in the codebook detailing how measurement equivalence was established, there needs to be information that clearly states how the sample was obtained; how the data were collected; and a rationale for why the measures were selected. The procedures used to obtain the sample should be clearly described so that researchers using the dataset can be assured that their groups were sampled in the same manner. Different sampling strategies can result in certain groups being selected more than others to be in the dataset. Therefore, details about how the data were collected from each group need to be outlined, so that researchers know that procedural equivalence has been established. A rationale for why the measures were selected should be presented. The rationale should mention that the measures were selected based on the empirically based literature indicating that the measures have the same factor structure for the groups included in the sample. If the measures are standardized, the norms for each group should be reported. Cutoff scores for each group should be presented as well, so that researchers can determine if scalar equivalence has been established. The mode of data collection should be noted, as research has provided evidence that different modes of data collection are associated with different response styles, affect the quality of the data, and influence the way people respond.

We have identified a few of the obstacles that exist when one is conducting research to establish equivalence using national datasets, and we have provided recommendations to address these challenges. We realize that our recommendations cannot be implemented unless there are policies and procedures put into place. Policies and procedures have been developed to encourage researchers to share their data with the scientific community for secondary data analysis. These policies clearly state what needs to be in shared datasets. We are recommending that these policies be modified so that researchers are expected to include in their datasets what we have discussed above.

FUTURE DIRECTIONS

Social Work Doctoral Education

Social work researchers who are proficient in conducting research with diverse groups are capable of producing valid and reliable findings that significantly contribute to the empirically based literature. Therefore, it is important that schools of social work increase the number of proficient researchers. One way to do this is to introduce content in the doctoral research methodology courses that emphasizes the need to establish equivalence at all phases of the research process. This book can be used as a supplement to current textbooks that are used in these courses. Structural Equation Modeling (SEM) and the use of factor analysis (FA) are commonly taught in social work doctoral programs. The examples in this book could easily be used when SEM and FA are being discussed. Presently there is concern among statisticians that applied researchers are not keeping up with the advances in the development of the SEM and CFA procedures (Muthén & Asparouhov, 2012; Rindskopf, 2012). To address this criticism, social work doctoral students could be introduced to Multiple Group Bayesian Confirmatory Factor Analysis and Restricted Factor Analysis/Latent Moderated Structure in their SEM course. These two techniques can be used to address some of the complexities associated with diversity.

It is our hope that this book will have an influence in shaping what social work doctoral students learn about conducting research with diverse groups. We also hope that this book encourages doctoral students, social work researchers, and social work educators to further examine all of the various statistical procedures that can be used to assess measurement and structural equivalence the ones discussed in this text, and others that were beyond the scope of this book. Moreover, we hope this book will encourage them to routinely test for measurement equivalence when conducting studies that look at group differences. It is our hope that our book will heighten awareness of the need to test for measurement and structural equivalence when conducting research with diverse groups and that establishing equivalence at all the phases of the research process is important in producing methodologically sound and valuable research that can lead to effective interventions and public policies.

In conclusion, we hope we have provided our readers with insight into the processes and procedures that need to be used in order to achieve research-design equivalence. Our review of current research methodology books revealed that the discussion of issues related to conducting research with diverse groups is often neglected or is limited at best. In situations where conducting research with diverse groups has been discussed, the focus has been on methods related to the implementation of the study (e.g., recruitment and retention of participants from diverse groups, and the problem-formulation phase of the research process).

Additionally, there is limited discussion in statistics books on challenges related to analyzing data with diverse groups. One important contribution of this book is that it has content about such research methods and statistics in one resource. An unique aspect of this book is that it discusses the need to establish equivalence at each phase of the research process and describes the processes and procedures that are used to do this. Finally, this book has heightened one's awareness of the need take into consideration contextual factors when establishing measurement equivalence. Processes and procedures that achieve research-design equivalence are important for researchers to implement, regardless of what phenomenon one is studying using diverse groups. Therefore, we feel that what has been presented in this book can advance the field of research in several disciplines in addition to social work.

Appendix A

Chi-Square Difference Testing Using the Satorra-Bentler Scaled Chi-Square: Hispanic and African American Males

Dr. Scott Colwell from the University of Guelph has developed a web-based calculator that will perform these calculations (See http://www.uoguelph.ca/~scolwell/qmlinks2.html). The calculations and instructions presented are based on the information found on the Mplus website (http://statmodel.com/chidiff.shtml).

1. Difference in test scaling correction cd.

cd = ((d0 * c0) – (d1 * c1))/(d0 – d1); d0 = df in the nested model. c0 = scaling factor for the nested model. d1 = df in the comparisons model, and c1 = scaling correction for the comparison model.

$$cd = ((34 * 1.414) - (32 * 1.374))/(34 - 32) = 2.054$$

2. Compute the MLR scaled chi-square difference test.

T0 = MLR chi-square values for the nested model. T1 = MLR chi-square values for the comparison model.

$$\Delta\text{MLR}\ \chi^2_{(\Delta df)} = ((T0 * c0) - (T1 * c1))/cd$$
$$\Delta\text{MLR}\ \chi^2_{(2)} = ((82.30 * 1.414) - (56.44 * 1.374))/2.054$$
$$\Delta\text{MLR}\ \chi^2_{(2)} = 18.90,\ p < .01$$

ESTABLISHING THE BASE CFA MODEL HISPANIC AND AFRICAN AMERICAN FEMALES

1. Difference in test scaling correction cd.

$$cd = ((34 * 1.281) - (31 * 1.22))/(34 - 31) = 1.92$$

2. Compute the MLR scaled chi-square difference test.

T0 = MLR chi-square values for the nested model. T1 = MLR chi-square values for the comparison model.

$$\Delta\text{MLR}\ \chi^2_{(\Delta Newdf)} = ((T0 * c0) - (T1 * c1))/cd$$
$$\Delta\text{MLR}\ \chi^2_{(3)} = ((91.98 * 1.281) - (42.68 * 1.22))/1.92$$
$$\Delta\text{MLR}\ \chi^2_{(3)} = 34.25,\ p < .01$$

Appendix B

Adjusted Chi-Square Difference Test: Configural versus Weak Factor Equivalence Model

In this example the nested model is the weak factor equivalence model ($\chi^2_{(73)}$ = 118.19, Scale Correction factor = 1.32 and the configuralmodel ($\chi^2_{(64)}$ = 118.29, Scale Correction factor = 1.31) is the comparison model

1. Difference in test scaling correction cd.

   cd = ((d0 * c0) – (d1*c1)) / (d0 – d1)
   cd = ((d0 * c0) – (d1*c1)) / (d0 – d1);
   cd = ((73*1.308) – (64*1.322))/(73 – 64);
   cd = 1.21

      d0 = df in the nested model
      c0 = scaling factor for the nested model.
      d1= df in the comparisons model.
      c1 = scaling correction for the comparison model

2. Compute the MLR scaled chi-square difference test ΔMLR $\chi^2_{(9)}$.

$$\Delta \text{MLR } \chi^2_{(\Delta \text{Newdf})} = ((\text{T0*c0}) - (\text{T1 *c1}))/\text{cd}$$
$$\Delta \text{MLR } \chi^2_{(9)} = ((118.19*1.31) - (118.59*1.32))/1.21$$
$$\Delta \text{MLR } \chi^2_{(9)} = 1.42$$

T0 = MLR chi-square values for the nested model.
T1 = MLR chi-square values for the comparison model.

Appendix C

Structural Equation Modeling Programs for Conducting Measurement Equivalency Analyses

Statistical Programs	Resources
AMOS	(B. Byrne, 2009)
EQS	(B. M. Byrne, 2006)
LISREL	(Raykov, 2004)
SAS PROC CALIS	(Jones-Farmer, Pitts, & Rainer, 2008)
STATA	(Gregorich, 2006)

References

Ackerman, T. A. (1992). A didactic explanation of item bias, item impact, and item validity from a multidimensional perspective. *Journal of Educational Measurement, 29*(1), 67–91.

Adamsons, K., & Buehler, C. (2007). Mothering versus fathering versus parenting: Measurement equivalence in parenting measures. *Parenting Science & Practice, 7*, 271–303. doi:10.1080/15295190701498686

Alfredo, J., Rueda, R., Salazar, J. J., & Higareda, I. (2005). Within-group diversity in minority disproportionate representation: English language learners in urban school districts. *Exceptional Children, 71*, 283–300.

Ashton, M. C., & Paunonen, S. V. (1998). The structured assessment of personality across cultures. *Journal of Cross-Cultural Psychology, 29*, 150–172.

Baraldi, A. N., & Enders, C. K. (2010). An introduction to modern missing data analyses. *Journal of School Psychology, 48*, 5–47. doi:10.1016/j.jsp.2009.10.001

Barnette, J. J. (2000). Effects of STEM and Likert response option reversals on survey internal consistency: If you feel the need, there is a better alternative to using those negatively worded STEMs. *Educational & Psychological Measurement, 60*, 361–370. doi:10.1177/00131640021970592

Baser, O. (2006). Too much ado about propensity score models? Comparing matching methods of propensity score matching. *Value in Health, 9*, 377–385. doi:10.1111/j.1524-4733.2006.00130.x

Baumgartner, H., & Steenkamp, J. E. M. (2001) Response styles in marketing research: A cross-national investigation. *Journal of Marketing Research, 38*, 143–156.

Berg, G. D., Johnson, A., & Fleegler, E. (2003). Clinical and utilization outcomes for a pediatric and adolescent telephonic asthma care support program: A propensity score-matched cohort study. *Disease Management & Health Outcomes, 1*, 737–743.

Bentler, P., & Yuan, K. H. (1999). On adding a mean structure to a covariance structure model. *Department of Statistics Papers*, 1–18. Retrieved on August 8, 2013 from http://repositories.cdlib.org/cgi/viewcontent.cgi?article=1179&context=uclastat

Bernal, G., & Sharron-del-Rio, M. G. (2001). Are empirically supported treatments valid for ethnic minorities? Toward an alternative approach for treatment research. *Cultural Diversity & Ethnic Minority Psychology, 7*, 328–342. doi:10.1037//1099-9809.7.4.328

Bobko, P., Roth, P. L., & Bobko, C. (2001). Correcting the effect size of *d* for range restriction and unreliability. *Organizational Research Methods, 4*(1), 46–61. doi: 10.1177/109442810141003

Bollen, K. A. (1989). *Structural Equations with Latent Variables*. New York: Wiley-Interscience.

Bollen, K. A. (2002). Latent variables in psychology and the social sciences. *Annual Review of Psychology, 53*(1), 605–634.

Borsboom, D. (2006). When does measurement equivalence matter? *Medical Care, 44*(Supplement 3), S176–S181. doi:10.1097/01.mlr.0000245143.08679.cc

Bowen, D. J, Bradford, J., & Powers, D. (2007). Comparing sexual minority status across sampling methods and populations. *Women & Health, 44*, 121–134. doi:10.1300/J013v44n02_07

Bowling, A. (2005). Mode of questionnaire administration can have serious effects on data quality. *Journal of Public Health, 27*, 281–291. doi:10.1093/pubmed/fdi031

Byrant-Bedell, K., & Waite, R. (2010). Understanding major depressive disorder among middle-age African American men. *Journal of Advanced Nursing, 66*, 2050–2060. doi:10.1111/j.1365-2648.2010.05345.x

Byrne, B. M. (2004). Testing for multigroup invariance using AMOS graphics: A road less traveled. *Structural Equation Modeling: A Multidisciplinary Journal, 11*(2), 272–300. doi: 10.1207/s15328007sem1102_8

Byrne, B. M. (2006). *Structural Equation Modeling with EQS*. New York: Taylor & Francis Group.

Byrne, B. M. (2008). Testing for multigroup equivalence of a measuring instrument: A walk through the process. *Psicothema, 20*(4), 872–882.

Byrne, B. M. (2009). *Structural Equation Modeling with AMOS: Basic Concepts, Applications, and Programming* (2nd ed.). New York: Taylor and Francis Group.

Byrne, B. M. (2011). *Structural equation modeling with Mplus 7.0s: Basic concepts, applications, and programming*. Thousands Oaks, CA: Sage Publications.

Byrne, B. M. (2012). *Structural equation modeling with Mplus: Basic concepts, applications, and programming*. New York: Routledge.

Byrne, B. M., & Campbell, T. L. (1999). Cross-cultural comparisons and the presumption of equivalent measurement and theoretical structure: A look beneath the surface. *Journal of Cross-Cultural Psychology, 30*, 555–574. doi:10.1177/0022022199030005001

Byrne, B. M., Shavelson, R. J., & Muthén, B. (1989). Testing for the equivalence of factor covariance and mean structures: The issue of partial measurement equivalence. *Psychological Bulletin, 105*(3), 456–466. doi:10.1037/0033-2909.105.3.456

Byrne, B. M., & van de Vijver, F. (2010). Testing for measurement and structural equivalence in large-scale cross-cultural studies: Addressing the issue of nonequivalence. *International Journal of Testing, 10*, 107–132. doi:10.1080/15305051003637306

Byrne, B. M., & Stewart, S. M. (2006). TEACHER'S CORNER: The MACS Approach to Testing for Multigroup Invariance of a Second-Order Structure: A Walk Through the Process. *Structural Equation Modeling, 13*(2), 287–321.

Byrne, B. M., & Watkins, D. (2003). The issue of measurement equivalence revisited. *Journal of Cross-Cultural Psychology, 34*, 155–175. doi:10.1177/0022022102250225

Carmines, E. G., & Zeller, R. A. (1979). *Reliability and Validity Assessment.* Sage Publications Thousand Oaks, CA.

Carter-Black, J. D., & Kayama, M. (2011). Jim Crow's daughters different social class-different experience with racism. *AFFILIA: Journal of Women & Social Work, 26*(2), 169–181.

Chen, Y., Rendina-Gobioff, G., & Dedrick, R. F. (2010). Factorial invariance of a Chinese self-esteem scale for third and sixth grade students: Evaluating method effects associated with positively and negatively worded items. *International Journal of Educational & Psychological Assessment, 6*, 21–35.

Cheung, G. W., & Rensvold, R. B. (2000). Assessing extreme and acquiescence response sets in cross-cultural research using structural equation modeling. *Journal of Cross-Cultural Psychology, 31*, 187–212. doi:10.1177/0022022100031002003

Clogg, C. C. (1995). Latent class models. In G. Arminger, C. C. Clogg, & M. E. Sobel (eds.), *Handbook of statistical modeling for the social and behavioral sciences* (pp. 311–359). New York: Plenum Press.

Crockett, L. J., Randall, B. A., Shen, Y., Russell, S. T., & Driscoll, A. K. (2005). Measurement equivalence of the Center for Epidemiological Studies Depression Scale for Latino and Anglo adolescents: A national study. *Journal of Consulting & Clinical Psychology, 73*, 47–58. doi:10.1037/0022-006X.73.1.47

Cronbach, L. J. (1950). Further evidence on response sets and test design. *Educational & Psychological Measurement, 10*, 3–31.

DeShon, R. P. (2004). Measures are not invariant across groups without error variance homogeneity. *Psychology Science, 46*, 137–149.

DiStefano, C., & Motl, R. W., (2006). Further investigating method effects associated with negatively worded items on self-report surveys. *Structural Equation Modeling, 13*, 440–464.

Elliott, M. N., Zaslavsky, A. M., Goldstein, E., Lehrman, W., Hambarsoomians, K., & Beckett, M. K., et al. (2008). Effects of survey mode, patient mix, and nonresponse on CAHPS hospital survey scores. *Health Services Research, 44* (part 1), 501–518. doi:10.1111/j.1475-6773.2008.00914.x

Enders, C. (2006). A primer on the use of modern missing-data methods in psychosomatic medicine research. *Psychosomatic Medicine, 68*, 427–436. doi: 10.1097/01.psy.0000221275.75056.d8

Farmer, G. L. (2002). The dimensionality of youths' oriented interests. *Journal of Social Service Research, 29*(2), 1–38.

Fleishman, J. A. (2004). Using MIMIC models to assess the influence of differential item functioning. Paper presented at the Conference on Improving Health Outcomes Assessment Based on Modern Measurement Theory and Computerized Adaptive Testing, Bethesda (Maryland) Hyatt, June 24.

Fleishman, J. A, & Benson, J. (1987). Using LISREL to evaluate measurement models and scale reliability. *Educational & Psychological Measurement, 47*(4), 925–939. doi:10.1177/0013164487474008

Gonzalez, R., & Griffin, D. (2001). Testing parameters in structural equation modeling: Every "one" matters. *Psychological Methods, 6*, 258–269.

Graham, J. W. (2009). Missing data analysis: Making it work in the real world. *Annual Review of Psychology, 60*, 549–576. doi: 10.1146/annurev.psych.58.110405.085530

Gregorich, S. E. (2006). Do self-report instruments allow meaningful comparisons across diverse population groups? Testing measurement equivalence using the confirmatory analysis framework. *Medical Care, 44*, S78–S94. doi:10.1097/01.mlr.0000245454.12228.8f

Guo, S., Barth, R. P., & Gibbons, C. (2006). Propensity score matching strategies for evaluating substance abuse services for child welfare clients. *Children & Youth Services Review, 28*, 357–383. doi:10.1016/j.childyouth.2005.04.012

Hayduk, L. A., & Glaser, D. N. (2000). Jiving the four-step, waltzing around factor analysis, and other serious fun. *Structural Equation Modeling, 7*(1), 1–35.

Ibrahim, A. M. (2001). Differential responding to positive and negative items. The use of a negative item in a questionnaire for course and faculty evaluation. *Psychological Reports, 88*, 497–500.

Ibrahim, S. A., Scott, F. E., Cole, D. C., Shannon, H. S., & Eyles, J. (2001). Job strain and self-reported health among working women and men: An analysis of the 1994–1995 Canadian National Population Health Survey. *Women's Health, 33*, 105–124.

Johnson, T. P. (2006). Methods and frameworks for cross-cultural measurement. *Medical Care, 44*, S17–S20. doi:10.1097/01.mlr.0000245424.16482.f1

Johnson, T. P., Shavitt, S., & Holbrook, A. L. (2011). Survey response styles across cultures. In D. Matsumoto & F. J. R. van de Vijver (eds.). *Cross-Cultural research methods in psychology.* (pp. 130–176). New York: Cambridge University Press.

Jones-Farmer, L. A., Pitts, J. P., & Rainer, R. K. (2008). A note on multigroup comparisons using SAS PROC CALIS. *Structural Equation Modeling, 15*(1), 154–173. doi: 10.1080/10705510701758414

Jordan. L. A., Marcus, A. C., & Reeder, L. G. (1980). Response styles in telephone and household interviewing: A field experiment. *Public Opinion Quarterly, 44*, 210–222.

Jöreskog, K. (1971). Simultaneous factor analysis in several populations. *Psychometrika, 36*, 409–426.

Jöreskog, K. G. (1993). Testing structural equation models. In K. A. Bollen & J. S. Long (eds.), *Testing structural equation models*, (pp. 294–316). Newbury Park, CA: Sage.

Kankaras, M., & Moors, G. (2011). Measurement equivalence and extreme response bias in comparison of attitudes across Europe: A multigroup latent class factor approach. *Methodology: European Journal of Research Methods for the Behavioral & Social Sciences, 7*, 68–80. doi:10.1027/1614-2241/a000024

Kingery, J. N., Ginsberg, G. S., & Burstein, M. (2009). Factor structure and psychometric properties of the Multidimensional Anxiety Scale for children in an African American adolescent sample. *Child Psychiatry & Human Development, 40*, 287–300. doi: 10.1007/s/0578-009-0126-0.

Klem, L. (2000). Structural equation modeling. In L. G. Grimm & P. R. Yarnold (eds.), *Reading and understanding more multivariate statistics* (pp. 227–260). Washington, DC: American Psychological Association.

Knight, G. P., Roosa, M. W., & Umaña-Taylor, A. J. (2009). *Studying ethnic minority and economically disadvantaged populations: Methodological challenges and best practices.* Washington, DC: American Psychological Association.

Knight, G. P., Virdin, L. M., Ocampo, K. A., & Roosa, M. (1994). An examination of the cross-ethnic equivalence of measures of negative life events and mental health among Hispanic and Anglo-American children. *American Journal of Community Psychology, 22*, 767–783.

Kriston, L. (2013). Dealing with clinical heterogeneity in meta-analysis. Assumptions, methods, interpretation. *International Journal of Methods in Psychiatric Research, 22*(1), 1–15.

Kumar, V. (2000). *International marketing research.* Upper Saddle River, NJ: Prentice-Hall.

Leung, K., & van de Vijver, F. J. R. (2008). Strategies for strengthening causal inferences in cross cultural research: Consilience approach. *International Journal of Cross Cultural Management, 8*, 145–169. doi:10.1177/1470595808091788

Liamputtong, P. (2010). *Performing qualitative cross-Cultural research.* Cambridge, United Kingdom: Cambridge University Press.

Lubke, G. H., Dolan, C. V., Kelderman, H., & Mellenbergh, G. J. (2003). On the relationship between sources of within- and between-group differences and measurement equivalence in the common factor model. *Intelligence, 31*(6), 543–566. doi: 10.1016/s0160-2896(03)00051-5

Marin, G., Gamba, T. J., & Marin, B. V. (1992). Extreme response style and acquiescence among Hispanics: The role of acculturation and education. *Journal of Cross-Cultural Psychology, 23,* 498–509. doi: 10.1177/0022022192234006

Marsh, H. W., Scalas, L. F., & Nagengast, B. (2010). Longitudinal tests of competing factor structures for Rosenberg Self-Esteem Scale: Traits, ephemeral artifacts, and stable response styles. *Psychological Assessment, 22,* 366–381. doi:10.1037/a0019225

Matsumoto, D., & van de Vijver, F. J. R. (2011). *Cross-Cultural research methods in psychology.* New York: Cambridge University Press.

McCabe, S. E., Hughes, T. L., Bostwick, W., Morales, M., & Boyd, C. J. (2012). Measurement of sexual identity in surveys: Implications for substance abuse research. *Archives of Sexual Behavior, 41*(3), 649–657. doi: 10.1007/s10508-011-9768-7

Meade, A. W., & Bauer, D. J. (2007). Power and precision in confirmatory factor analytic tests of measurement equivalence. *Structural Equation Modeling, 14,* 611–635.

Meredith, W. P. (1993). Measurement equivalence, factor analysis and factorial invariance. *Psychometrika, 58,* 525–543. doi:10.1007/BF02294825

Meredith, W. P., & Teresi, J. A. (2006). An essay on measurement and factorial invariance. *Medical Care, 44*(11), S69–S77.

Milfont, T. L., & Fischer, R. (2010). Testing measurement equivalence across groups: Applications in cross-cultural research. *International Journal of Psychological Research, 3*(1), 111–121.

Millsap, R. E. (1989). Sampling variance in the correlation coefficient under range restriction: A Monte Carlo study. *Journal of Applied Psychology, 74*(3), 456–461. doi: 10.1037/0021-9010.74.3.456

Millsap, R. E. (2007). Structural equation modeling made difficult. *Personality & Individual Differences, 42*(5), 875–881. doi: 10.1016/j.paid.2006.09.021

Millsap, R. E., & Everson, H. T. (1993). Methodology review: Statistical approaches for assessing measurement bias. *Applied Psychological Measurement, 17*(4), 297–334.

Moisio, P. (2004). A latent class application to the multidimensional measurement of poverty. *Quality & Quantity, 38*(6), 703–717.

Muthén, B. (2002). Beyond SEM: General latent variable modeling. *Behaviormetrika, 29*(1), 81–117.

Moors, G. (2004). Facts and artefacts in the comparison of attitudes among ethnic minorities. A multi-group latent class structure model with adjustment for response style behaviour. *European Sociological Review, 20*, 303–320. doi:10.1093/esr/jch026

Muthén, B. O. (1989). Latent variable modeling in heterogeneous populations. *Psychometrika, 54*, 557–585.

Muthén, B. O. (2002). Beyond SEM: General latent variable modeling. *Behaviormetrika, 29*(1), 81–117.

Muthén, B. O. (2012). Chi-square difference test using the Satorra-Bentler scaled chi-square. Retrieved Nov. 11, 2012, from http://www.statmodel.com/chidiff.shtml.

Muthén, B., & Asparouhov, T. (2012). Bayesian structural equation modeling: A more flexible representation of substantive theory. *Psychological Methods, 17*(3), 313–335. doi: 10.1037/a0026802

Muthén, L., & Muthén, B. (1998–2007). *Mplus User's Guide* (4th ed.). Los Angeles: Muthén & Muthén.

Neely-Barnes, S. (2010). Latent class models in social work. *Social Work Research, 34*, 114–121.

Nimon, K., & Reio Jr., T. G. (2011). Measurement equivalence: A foundational principle for quantitative theory building. *Human Resource Development Review, 10*, 198–214. doi:10.1177/1534484311399731

Nugent, W. R. (2012). The interchangeability of scores from different measures and meta-analytic effect size comparability. *Journal of the Society for Social Work & Research, 3*(4), 319–336. doi: 10.5243/jsswr.2012.14

Nugent, W. R., & Hankins, J. A. (1992). A comparison of classical, item response, and generalizability theories of measurement. *Journal of Social Service Research, 16*(1–2), 11–39.

Orhede, E., & Kreiner, S. (2000). Item bias in indices measuring psychosocial work environment and health. *Scandinavian Journal of Work, Environment, & Health, 26*, 263–272. doi:10.5271/sjweh.541

Pett, M. A., Lackey, N. R., & Sullivan, J. J. (2003). *Making Sense of Factor Analysis.* Thousand Oaks, CA: Sage Publication.

Ployhart, R. E., & Oswald, F. L. (2004). Applications of mean and covariance structure analysis: Integrating correlational and experimental approaches. *Organizational Research Methods, 7*(1), 27–65. doi:10.1177/1094428103259554

Polsky, D., Eremina, D., Hess, G., Hill, J., Hulnick, S., Roumm, A.... Kallich J. (2009). The importance of clinical variables in comparative analyses using propensity-score matching: The case of ESA costs for the treatment of chemotherapy-induced anaemia. *Pharmacoeconomics, 27*(9), 755–765.

Ray, J. J. (1983). Reviving the problem of acquiescent response bias. *Journal of Social Psychology, 121*, 81–96.

Raykov, T. (2004). Behavioral scale reliability and measurement equivalence evaluation using latent variable modeling. *Behavior Therapy*, *35*(2), 299–331. doi: 10.1016/S0005-7894(04)80041-8

Reise, S. P., & Haviland, M. G. (2005). Item response theory and the measurement of clinical change. *Journal of Personality Assessment*, *84*, 228–238. doi:10.1207/s15327752jpa8403_02

Rindskopf, D. (2012). Next steps in Bayesian structural equation models: Comments on, variations of, and extensions to Muthén and Asparouhov (2012). *Psychological Methods*, *17*(3), 336–339. doi: 10.1037/a0027130

Rosato, N. S., & Baer, J. C. (2012). Latent-class analysis: A method for capturing heterogeneity. *Social Work Research*, *36*, 61–69. doi:10.1093/swr/svs006

Saris, W. E., Satorra, A., & Sorbom, D. (1987). The detection and correction of specification errors in structural equation models. *Sociological Methodology*, *17*, 105–129.

Satorra, A. (2000). Scaled and adjusted restricted tests in multi-sample analysis of moment structures. In E. D. H. Heijmans, D. S. G. Pollock, & A. Satorra (eds.), *Innovations in multivariate statistical analysis* (pp. 233–247). London: Kluwer Academic Publishers.

Satorra, A., & Bentler, P. M. (2010). Ensuring positiveness of the scaled difference chi-square test statistic. *Psychometrika*, *75*(2), 243–248. doi: 10.1007/s11336-009-9135-y

Schaefer, J. L., & Graham, J. W. (2002). Missing data: Our view of the state of the art. *Psychological Methods*, *7*, 147–177. doi:10.1037/1082-989X.7.2.147

Schaffer, B. S., & Riordan, C. M. (2003). A review of cross-cultural methodologies for organizational research: A best-practices approach. *Organizational Research Methods*, *6*, 169–215. doi:10.1177/1094428103251542

Schick, V., Rosenberger, J. G., Herbenick, D., Calabrese, S. K., & Reece, M. (2012). Bidentity: Sexual behavior/identity congruence and women's sexual, physical and mental well-being. *Journal of Bisexuality*, *12*(2), 178–197. doi: 10.1080/15299716.2012.674855

Sean, J. (2005). Standing in the shadow: Understanding and overcoming depression in black men. *International Journal of Men's Health*, *4*, 93–94.

Sharma, S., Durvasula, S., & Ployhart, R. E. (2011). The analysis of mean differences using mean and covariance structure analysis: Effect size estimation and error rates. *Organizational Research Methods*, *15*, 75–102. doi: 10.1177/1094428111403154

Smith, J., Todd, P. (2005). Does matching overcome Lalonde's critique of non-experimental estimators? *Journal of Econometrics*, *125*, 305–353. doi:10.1016/j.jeconom.2004.04.011

Sorbom, D. (1989). Model modification. *Psychometrika*, *54*, 371–384.

Spini, D. (2003). Measurement equivalence of 10 value types from the Schwartz Value Survey across 21 countries. *Journal of Cross-Cultural Psychology*, *34*(1), 3–23.

Steinmetz, H., Schwens, C., Wehner, M. C., & Kabst, R. (2011). Conceptual and methodological issues in comparative HRM research: The Cranet Project. *Human Resources Management Review, 21*, 16–26.

Stricker, L. J. (1963). Acquiescence and social desirability response styles, item characteristics, and conformity. *Psychological Reports, 12*, 319–341.

Supple, A. J., & Plunkett, S. W. (2011). Dimensionality and validity of the Rosenberg Self-Esteem Scale for use with Latino adolescents. *Hispanic Journal of Behavioral Science, 33*, 39–53. doi:10.1177/0739986310387275

Tenijenhuis, J., Tolboom, E., Resing, W., & Bleichrodt, N. (2004). Does cultural background influence the intellectual performance of children from immigrant groups? The RAKIT intellectual test for immigrant children. *European Journal of Psychological Assessment, 20*, 10–26.

Teresi, J. A. (2006). Overview of quantitative measurement methods: Equivalence, invariance, and differential item functioning in health applications. *Medical Care, 44*, S39–S49.

Thyer, B. (2010). *Handbook of Social Work Research* (2nd ed.). Thousand Oaks, CA: Sage Publications.

Tourangeau, R., & Smith, T. W. (1996). Asking sensitive questions. The impact of data collection mode, question format, and question context. *Public Opinion Quarterly, 60*, 275–304. doi:275-304 doi:10.1086/297751

Tran, T. V. (2009). *Developing cross-Cultural measurement*. New York: Oxford Press.

U.S. Census Bureau (2012). *American Community Survey for One Year*. Washington, DC: U.S. Department of Commerce and Economic Statistics Administration. Retrieved on July 21, 2013, from http://censuschannel.net/cc/news/2010-median-household-individual-income-asian-americans-top-the-list-1330.

U.S. Census Bureau (2011a). *The Black population: 2010: 2010 Census Briefs*. Washington, DC: U.S. Department of Commerce and Economic Statistics Administration.

U.S. Census Bureau (2011b). *The Hispanic population: 2010: 2010 Census Briefs*. Washington, DC: U.S. Department of Commerce and Economic Statistics Administration.

U.S. Census Bureau (2012). *Census Bureau Table 6: Percent of the projected population by race and Hispanic origins for the United States: 2015 to 2060* (NP2012-T6). Washington, DC: U.S. Department of Commerce and Economic Statistics Administration. Retrieved on July 21, 2013, from http://www.census.gov/population/projections/data/national/2012/summarytables.html.

U.S. Department of the Interior. (2013). *Native Americans*. Retrieved August 8, 2013, from http://www.doi.gov/library/internet/native.cfm.

van de Vijver, F., & Leung, K. (1997). *Methods and data analysis for cross-Cultural research*. Thousand Oaks, CA; Sage.

Van Herk, H. Poortinga, Y. H., & Verhallen, T. M. M. (2004). Response styles in rating scales—Evidence of method bias in data from six EC countries. *Journal of Cross-Cultural Psychology, 35*, 346–360. doi:10.1177/0022022104264126

Van Herk, H., Poortinga, Y. H., & Verhallen, T. M. M. (2005). Equivalence of survey data: Relevance for international marketing. *European Journal of Marketing, 39*, 351–364. doi:10.1108/03090560510581818

Vandenberg, R. J., & Lance, C. E. (2000). A review and synthesis of the measurement equivalence literature: Suggestions, practices, and recommendations for organizational research. *Organizational Research Methods, 3*, 4–70. doi:10.1177/109442810031002

Viswanathan, M. (2005). *Measurement error and research design*. Thousand Oaks: Sage Publications.

Weijters, B., Schillewaert, N., & Geuens, M. (2008). Assessing response styles across modes of data collection. *Journal of the Academy of Marketing Science, 36*, 409–422. doi:10.1007/S11747-007-0077-6

Wicherts, J. M., & Dolan, C. V. (2010). Measurement equivalence in confirmatory factor analysis: An illustration using IQ test performance of minorities. *Educational Measurement: Issues & Practice, 29*(3), 39–47. doi: 10.1111/j.1745-3992.2010.00182.x

Widaman, K. F., & Reise, S. P. (1997). Exploring the measurement equivalence of psychological instruments: Applications in the substance use domain. In K. J. Bryant, M. Windle, & S. G. West (eds.), *The science of prevention: Methodological advances from alcohol and substance abuse research* (pp. 281–324). Washington, DC: American Psychological Association.

Widhiarso, W. (2010). Estimate reliability measurement for multidimensional scales. *SSRN Electronic Journal*, 1–8. doi: 10.2139/ssrn.1597532

Willoughby, M. T., Wirth, R. J., & Blair, C. B. (2012). Executive function in early childhood: Longitudinal measurement equivalence and developmental change. *Psychological Assessment, 24*(2), 418–431. doi: 10.1037/a0025779

Wu, A. D., & Zumbo, B. D. (2007). Decoding the meaning of factorial invariance and updating the practice of multi-group confirmatory factor analysis: A demonstration with TIMSS data. *Practical Assessment, Research & Evaluation, 12*(3), 1–27.

Wu, P.-C. (2009). Differential functioning of the Chinese version of Beck Depression Inventory-II in adolescent gender groups: Use of a multiple-group mean and covariance structure model. *Social Indicators Research, 96*(3), 535–550. doi:10.1007/s11205-009-9491-0

Yanyun, Y., & Green, S. B. (2011). Coefficient Alpha: A reliability coefficient for the 21st century?. *Journal of Psychoeducational Assessment, 29*(4), 377–392. doi:10.1177/0734282911406668

Ye, C., Fulton, J., & Tourangeau, R. (2011). More positive or more extreme? A meta analysis of mode differences in response choice. *Public Opinion Quarterly, 75*, 349–365. doi:10.1093/poq/nfr009

Yoon, M., & Millsap, R. E. (2007). Detecting violation of factorial invariance using data-based specification searches: A Monte Carlo study. *Structural Equation Modeling, 14*(3), 435–463.

Yuan, K.-H., & Bentler, P. M. (2006). Mean comparison: Manifest variable versus latent variable. *Psychometrika, 71*(1), 139–159. doi: 10.1007/s11336-004-1181-x

Index